MONTGOMERY COUNTY LIBRARY
3 3543 01355 9091

D1154172

The Wonderful World of Relativity

The Wonderful World of Relativity

A precise guide for the general reader

Andrew M. Steane

OXFORD
UNIVERSITY PRESS

OXFORD
UNIVERSITY PRESS

Great Clarendon Street, Oxford OX2 6DP

Oxford University Press is a department of the University of Oxford.
It furthers the University's objective of excellence in research, scholarship,
and education by publishing worldwide in

Oxford New York

Auckland Cape Town Dar es Salaam Hong Kong Karachi
Kuala Lumpur Madrid Melbourne Mexico City Nairobi
New Delhi Shanghai Taipei Toronto

With offices in

Argentina Austria Brazil Chile Czech Republic France Greece
Guatemala Hungary Italy Japan Poland Portugal Singapore
South Korea Switzerland Thailand Turkey Ukraine Vietnam

Oxford is a registered trade mark of Oxford University Press
in the UK and in certain other countries

Published in the United States
by Oxford University Press Inc., New York

First published 2011

Reprinted 2012

British Library Cataloguing in Publication Data
Data available

Library of Congress Cataloging in Publication Data
Steane, Andrew M.
The wonderful world of relativity : a precise guide for
the general reader / Andrew M. Steane.
p. cm.
Includes index.
ISBN 978–0–19–969461–7
1. Relativity (Physics)—Popular works. I. Title.
QC173.57.S74 2011
530.11—dc23 2011026848

Typeset by SPI Publisher Services, Pondicherry, India
Printed in Great Britain
on acid-free paper by
CPI Group (UK) Ltd, Croydon, CR0 4YY

ISBN 978–0–19–969461–7

3 5 7 9 10 8 6 4 2

To Emma

Contents

1

Introduction

Relativity, or, to give it its full name, Einstein's Special Theory of Relativity, is one of the most fascinating and fun areas of modern physics. It involves some ideas which at first sight seem very strange and mind-bending, such as that space and time can, in a manner of speaking, change shape. However, it does not require advanced mathematics to obtain a good understanding of the main ideas of Special Relativity, and any inquisitive person who knows how to plot a graph and is not put off by a square-root sign can hope to understand it well. This is why this book is aimed at the pre-university age bracket, or at the general scientific reader who would like to have a fully correct treatment at an accessible mathematical level. I want to give to such eager minds something thorough and precisely correct, but also fun and engaging.

Relativity is a theory of space and time and motion. As usual in physics, the word 'theory' here means 'a thorough mathematical and physical description', not 'Einstein's opinion' or 'someone's idea'. What is rather wonderful about the theory of Relativity is that it makes the most astonishing claims that at first sight seem impossible, and yet it is so elegant and natural that by the time one has understood it those astonishing claims seem simply clear and beautiful.

For example, one claim is that if you take an ordinary wooden pole, a few metres long, and accelerate it gently up to very high speed, then it will shrink as it gets faster. It will shrink and shrink, down to a metre length, a few centimetres, a millimetre, and smaller still, as its speed approaches closer and closer to 299, 792, 458 metres per second. However, a person carrying the pole and moving along with it (perhaps they are both in a very fast space rocket) will not notice anything untoward: they will think the pole is just an ordinary wooden pole, a few metres long. As this person flies by planet Earth at high speed, viewers on Earth could take a look at life inside the rocket, and they would conclude that the reason the astronaut has not noticed anything is that he too has shrunk. But the shrinking occurs only along his direction of motion. As he stood or lay down his body shape would change in the most extraordinary way: Earth inhabitants would find him to be thin as a rake when lying down, but short and squat when standing up! (see figure 1.1). All his cells would change shape together, right down to their molecules and atoms, and atomic nuclei and elementary particles and fields. The whole rocket would be shrunk too. His life would also be proceeding in slow-motion. All his thoughts and actions, and all other movements inside the rocket, would be slow compared to their equivalents on Earth.

And yet the astronaut would be utterly unaware of any problem. As far as he is concerned, life on board the rocket is completely ordinary; it has its normal pace, and his body has its normal shape. He would go about his everyday life, listening to music and taking a drink of astro-coffee, just like he used to do on Earth. Such an astronaut would meanwhile look at planet Earth, and from his careful measurements as it whizzed by (remember he is in a fast spaceship) he would conclude that Earth is not approximately spherical, but squashed flat, and everything there is slowed down . . . even though we inhabitants find nothing unusual.

These statements seem very odd, but one must be careful not to present Relativity merely as a sort of freak show: 'look how amazing it is.' The aim of this book is that you should *understand*, in a clear and satisfying way, what is going on, so that instead of thinking 'amazing!' you will think, 'oh yes, it must be like that.' Surprise should give way to appreciation. Such understanding does not take away the wonder of the subject, but it gives to that

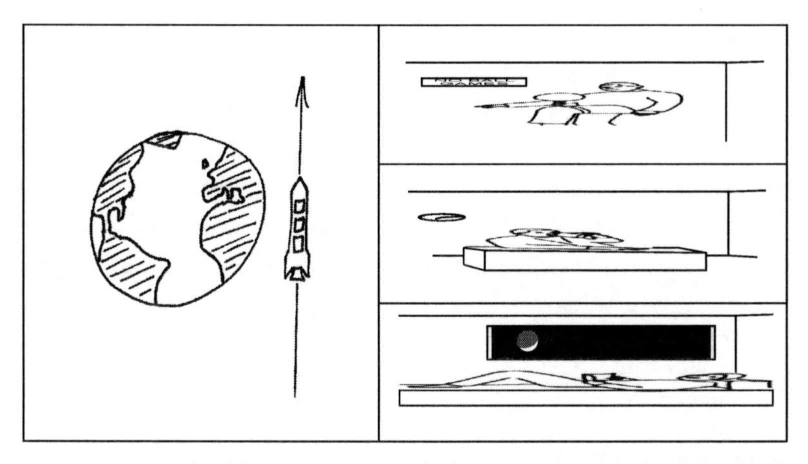

Figure 1.1: Strange behaviour of a fast astronaut. For a rocket travelling at very high speed past the Earth, the life of an astronaut on board seems very bizarre to inhabitants of Earth. If they observe the rocket by telescope, then they could create a feature film showing a day in the life of Robby the rocketeer. The editing team would first have to apply a correction to allow for the varying travel time of light from the rocket to the telescope. After doing this they would have a true record of the state of affairs from moment to moment in the rocket, according to an Earth-based definition of time. The images above show some frames from such a film. Robby is squashed flat! The squashing is always along his direction of motion, so he changes shape as he stands or sits or lies down. Also, his movements are very slow: his heart, lungs, brain, and all his cellular biology is about seven times slower than usual, as measured by Earth time. It takes a whole Earth week for his watch to register 24 hours. However, he appears completely at ease and unaware that anything strange is happening. Indeed, according to the theory of Relativity, from his point of view everything on board the rocket is normal, while Earth is slow and squashed to a pancake-like shape. By the end of this book, I hope not only to convince you of these things, but also to provide a sufficiently thorough understanding that they seem natural and expected! The amount of squashing and slowing depends on the speed of the rocket relative to Earth. The figure shows an illustrative case, where the speed is 297,000 kilometres per second. This is 99% of the speed of light.

wonder a different character, a deeper and richer flavour. Happy interest is turned into joyful recognition . . . or at least, that is the aim of my book.

The text assumes the reader has some general idea of Newton's Laws of Motion—for example, that a body will continue moving at constant speed in a straight line unless a force acts on it—and is willing to engage with some simple algebra. I expect this will either be readers who are at school and would like to get their teeth into some science that is accessible but complete and accurate, or other readers who have a general scientific interest. By the end of the book you will have a clear and accurate knowledge of the main principles of Relativity, of the nature of space and time, and of the equivalence of mass and energy.

Mathematics

Someone has said that for every equation you put in a science book, you have to reckon on halving the number of readers. Although this can be a useful warning to writers, I do not think the situation is quite like that. Mathematical equations are more like a language or a form of poetry: they will deter people who do not speak the language or understand the poem. What matters is whether the equations are understandable. Therefore, I shall say a little here about the mathematics in this book, to help you decide how you feel about this.

I guess most people have at least heard of the most famous equation in physics, $E = mc^2$. One of the main aims of this book is to help you understand the scientific ideas summed up by this equation, and how they were discovered. However, I hope that you are already comfortable with the idea that the equation is a form of shorthand. It states in brief that something indicated by the letter E (it is energy in fact) can be calculated by multiplying a quantity indicated by the letter m (mass) by the square of a quantity indicated by the letter c (the speed of light), and by the 'square' we mean the thing multiplied by itself: $c \times c$. The *physical meaning* of the equation requires a lot of discussion, which I will present in Chapter 10, but as far as the mathematics is concerned, it is quite simple: just some things multiplied together. In this book I am going to assume that you do not mind my using the shorthand 'mc^2' for 'mass multiplied by speed of light multiplied by speed of light'.

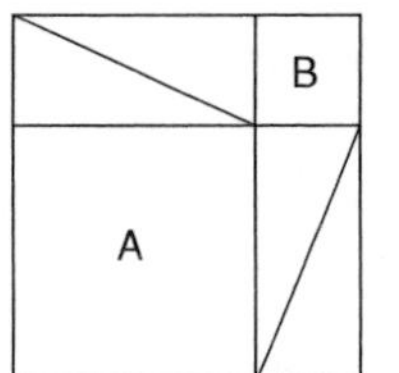

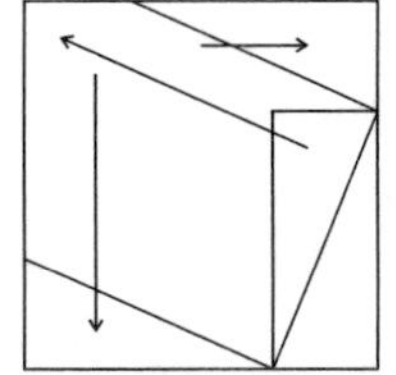
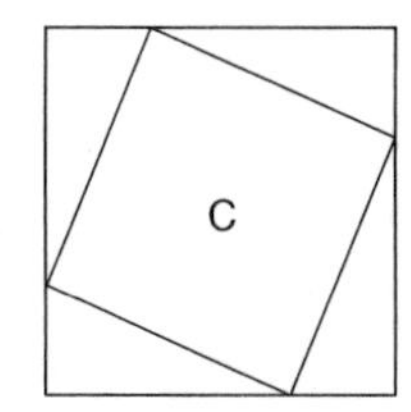

Figure 1.2: An elegant proof of Pythagoras's theorem. The left-hand diagram shows an outer square divided into two smaller squares and two rectangular regions. The rectangular regions have been cut into two triangles each. The four resulting triangles are all the same as each other ('congruent'). The two squares labelled A and B are clearly squares on two of the sides of one such triangle. The area of the whole figure is $A + B + 4a$, where a is the area of one triangle. The middle diagram shows how you can now slide the triangles to the corners, and thus create a diagram with symmetry under any rotation through a right angle. It follows that the region labelled C is another square, and it is clearly the square on the hypotenuse of the triangle. The area of the whole figure is unchanged by this rearrangement, so $A + B + 4a = C + 4a$, hence $A + B = C$, which is Pythagoras's theorem.

The letter c is almost always used for the speed of light. It was originally coined from the English word 'celerity' meaning swiftness, though that word has now largely fallen into disuse. Sometimes in science we like to use a Greek letter for one or more symbols, and I am going to use one in this book: it is γ, the Greek letter 'gamma'. There is nothing special about Greek letters. We could easily use a Roman letter such as g, but I am using γ for cultural reasons: it is the letter that the physics community has adopted for a factor called the 'Lorentz factor', which arises frequently in Relativity, and I think you might prefer to share in this practice in order to connect to that community and thus widen your cultural world a little.

Greek culture will also have an impact on our discussions through a famous theorem that we will need. This is Pythagoras's theorem concerning the sides of a right-angled triangle: 'the square on the hypotenuse is equal to the sum of the squares on the other two sides'. Figure 1.2 explains what this means, and also provides a beautiful proof. The 'hypotenuse' is the longest side and the 'square on the hypotenuse' is a geometric square placed against the hypotenuse so that the hypotenuse of the triangle forms one of its sides. When we say this square is 'equal to' the sum of the others (on the other two sides), we mean their areas are equal. Since the area of a square is obtained by multiplying the length of a side by itself, the theorem can also be stated as 'the square of the length of the hypotenuse is equal to the sum of the squares of the lengths of the other two sides.' That is, if the lengths of the sides are h, a, b, then $h \times h = a \times a + b \times b$, or more succinctly, $h^2 = a^2 + b^2$. This is how we mostly use the theorem nowadays.

It is also common to 'take the square root'—that is, to argue that if $h^2 = a^2 + b^2$ then it must follow that $h = \sqrt{a^2 + b^2}$, where the square-root sign $\sqrt{\ }$ has the meaning 'if x is any number, then $\sqrt{x}$ is the number that, when multiplied by itself, will give x'.

When we have two different quantities that are related to one another but are not the same, it is useful to use the same letter, but add a mark like an accent to one version and not to the other: for example, x and x', or y and y'. This is read either 'x and x-dash, y and y-dash' or 'x and x-prime, y and y-prime.' For example, we might use h to be the height of an aeroplane above the ground, and t to be the time of day, as indicated in the aeroplane's country of origin. We expect that as t increases, h first rises from zero to some thousands of feet, stays roughly constant, and finally falls (gently we hope) to zero again. We might also take an interest in the height of the aeroplane above sea level, and the time of day in the country of destination. We could use two more letters for those quantities—say d and u—but it can be awkward to find memorable letters, and we face a proliferation of symbols. In this case it can be more convenient to use the inverted comma ($'$), so the new quantities would be called h' and t'. For example, in this notation, we expect that h' differs from h by an amount giving the height of the terrain, and t' differs from t by a fixed number of hours.

Finally, we will need to perform some manipulations of equations. I will indicate the steps that are needed, but you are encouraged to work them through yourself and thus confirm to your own satisfaction that the answers I will give are correct. I hope you will find that rewarding, like solving a puzzle, with the added thrill that these are not man-made puzzles, but ones given to us by the subtle and beautiful patterns of the world we live in. I will also provide a few more puzzles and challenges as we go along.

2

A preview: the Laws of Motion

Before getting started on the more detailed reasoning of Relativity, it will be useful to introduce some terminology, and also to discuss a simple example where we see the theory of Special Relativity in action. The example is a particle moving in a straight line under the action of a constant force. Usually, books on Relativity do not introduce such an example until a careful discussion of space and time (or 'kinematics') has been presented. Here we will take a look at the example in order to have a foretaste of what is to come, and also to reassure you that although a thorough understanding of Relativity requires some subtle and (at first) difficult ideas, this does not mean that it completely changes everything we ever learned from Newton's Laws of Motion. In fact, those laws remain valid!

2.1 Relative and absolute

We have already mentioned that Relativity is called 'Special Relativity', which is simply to distinguish it from 'General Relativity'. General Relativity is Einstein's theory of gravitation and spacetime. It is very much more sophisticated, and it includes Special Relativity as a special case—hence the name. The relationship between them is similar to the relationship between geometry on an undulating surface and geometry on a flat plane.

The word 'relativity' is used because a central idea is that we have to consider carefully the difference between 'absolute' things, which do not depend on anyone's point of view, and 'relative' things, which might. For example, suppose I tell you that my house is on the left-hand side of the street. The problem with this statement, on its own, is that you cannot tell from it anything useful about the location of my house! This is because 'left' and 'right' are relative concepts: they depend on which way one is facing. To tell you something useful I would have to say something like 'if you face towards the park, with your feet on the ground, then my house will be on your left.'

Another example of a relative concept is *velocity*. I might say that 'my house is standing still.' I think you know what I mean by that, but, strictly speaking, the velocity of an object is a relative concept. After all, my house is standing on planet Earth, and the Earth is moving in its orbit around the Sun at approximately 67,000 miles per hour. To be more careful, I could say 'my house is not moving relative to planet Earth'.

An absolute property is one which does not depend on your point of view or your state of motion. Two examples are 'my house has a roof', and 'my house has 68 layers of bricks between the ground and the roof.' Here it does not matter whether I am speaking about the situation relative to the Earth or the Sun, nor in which direction one is facing: these statements remain true with no need to be clarified or modified. They are statements about absolute properties.

In Special Relativity we find that some things that we normally think of as absolute are in fact relative: they depend on motion. Not all things, but some. An example is the height of my house, as measured not by counting bricks, but by using a standard unit of measurement such as the metre. It turns out that the height measured in metres depends on the relative state of motion of the metre sticks and the house. The precise meaning of this statement will become clear as we go along.

You should note that the word 'relative' here does *not* mean 'a mere appearance' or 'illusory' or 'a matter of opinion'. If I say 'the road passes to the left of the tree' but someone else says 'the road passes to the right of the tree' it is not that the road merely 'appears to be' on one side or the other, nor that we should 'agree to differ'. No, in a case like this we should *not* agree to differ: we should figure out what is going on. We might learn, for example, that the road *really is* on the left (when you face north) and it *really is* on the right (when you face south). Similarly, suppose that a man walked down a carriage of a moving railway train, and the train was moving at 40 miles per hour past a station platform where I was seated. If someone says 'the man moved at a speed of 3 miles per hour' but I say 'the man moved at a speed of 43 miles per hour', then it is not a matter of mere appearances or opinions. Both statements concern concrete realities and both are correct; they are merely lacking in clarity of expression. A more careful statement might be 'the distance between the man and the end of the carriage changed at the rate of 3 miles per hour' or 'the distance between the man and the end of the platform changed at the rate of forty-three miles per hour.' Now both statements are clear, and both may be true without contradiction. In Special Relativity we would need to add further qualifying clauses, and further properties (not just speeds and directions) are found to depend on the state of motion of the object or body in question: in particular, properties such as length and time interval.

2.2 The laws of motion

In this book, Special Relativity will be introduced in two stages. First we will discuss 'kinematics', and then 'dynamics'. (The present chapter gives a preview of some aspects of dynamics). *Kinematics* is concerned with the basic concepts that are needed to discuss motion—concepts such as distance and time interval and velocity. It answers questions such as 'how can one spatial displacement be added to another?' The answer is that they add in the familiar way, like steps or arrows, that take into account both the length and the direction of each displacement. Mathematically, we say they add like 'vectors'. However, velocities add up in different way to this in Special Relativity. The question could be, for example, 'If a small meteor approaches Earth at velocity 100, 000 kilometres per second relative to Earth, and emits in the forward direction a fragment that moves with velocity 80, 000 kilometres per second relative to the meteor, then what is the velocity of the fragment relative to Earth?' The answer is surprising: it is not 180, 000 kilometres per second. It turns out that the velocities do *not* simply add up like vectors. Some further factors involving the speed of light have to be taken into account. We will see why in chapter 7.

Dynamics is concerned with what happens when physical entities push and pull on each other: it is the study of collisions and forces, and the motion that results from a force. In kinematics the main ideas are position, time, and velocity; in dynamics the main ideas are momentum, energy, force, and mass. In classical physics, that is, the physics that is valid when speeds are small compared to the speed of light and when quantum effects are negligible, the principles of dynamics are summarized by Newton's Laws of Motion:

1. If no forces act on a body, then it will move in a straight line at constant speed. In short, its velocity will be constant.
2. The rate of change of momentum of a body is equal to (has the same direction and the same size as) the total force on the body.
3. Whenever one body A exerts a force on another body B, the second body B exerts an equal and opposite force on body A.

Note that I have stated the second law in the best form 'rate of change of momentum' rather than 'mass times acceleration' (Newton equated the force to the rate of change of 'quantity of motion', and he had previously defined the 'quantity of motion' to be the product of mass and velocity, which is what we now call momentum). Also, I have stated the third law in a form that is clearer than the versions found in many textbooks.

Some textbooks on Relativity say that Newton's third law (on action and reaction) has to be abandoned. This is not true, but it is necessary in Relativity to apply the law correctly, which means *locally*. That is, the forces mentioned in the third law act *at the same point in space*. They act at the boundary between two different things, pushing them in opposite directions. One should not try to apply the third law to forces acting on bodies separated by some distance.

Now we can state a useful and interesting fact:

Useful Fact: *Newton's Laws of Motion remain correct in Special Relativity, as long as the relativistic equation for momentum is used.*

The 'relativistic equation for momentum' is the equation that the theory of Relativity introduces. It is:

$$\text{momentum} \quad \mathbf{p} = \frac{m\mathbf{v}}{\sqrt{1 - v^2/c^2}} \tag{2.1}$$

where v is the speed of the body, $\mathbf{v}$ is its velocity (the same size as the speed, but *velocity* takes the direction into account too) m is its mass (also called 'rest mass'; we will see why in Chapter 10) and c is the speed of light in vacuum. The symbol v^2 means 'v squared'—$v \times v$—and similarly for c^2, and the / sign means 'divided by' (like $\div$, but we prefer / because it reminds us of a fraction). We will find out where this formula comes from in chapter 10, but I want to introduce it now in order to give you a sense of where we are heading. The formula says that the momentum is calculated by taking the product $m\mathbf{v}$ of mass times velocity, and dividing it by a factor that relates to the speed of the object compared to the speed of light. In order to focus our thinking on the different parts of the formula, it is useful to introduce the notation

$$\gamma = \frac{1}{\sqrt{1 - v^2/c^2}} \tag{2.2}$$

(here is our friend γ (gamma), mentioned in the introduction), and then the formula for momentum can be written

$$\mathbf{p} = \gamma m\mathbf{v}$$

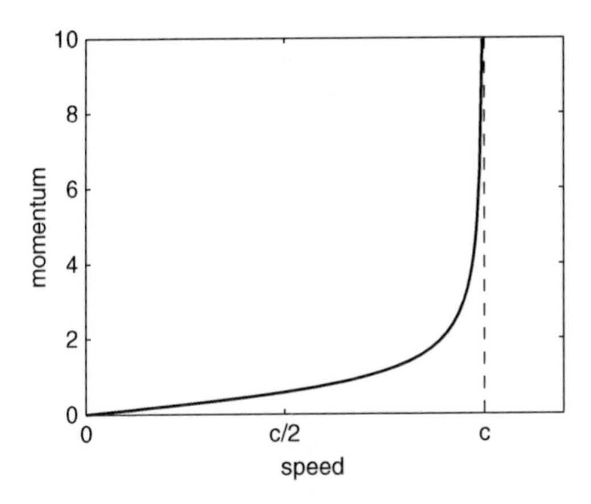

Figure 2.1: The relationship between momentum and speed, as predicted by Special Relativity. The momentum is shown in units of mc, where m is the rest mass. When the speed approaches the speed of light c, the momentum tends to infinity.

To understand the role of the Lorentz factor γ, note that if $v = 0$ then $\gamma = 1$, and if v is small compared to the speed of light c, then the fraction v^2/c^2 is very small. In this case $(1 - v^2/c^2)$ is as near to 1 as makes no difference, so then γ is almost exactly 1. Therefore, at low speed, momentum equals mass times velocity. This is the form that is familiar in pre-Relativity physics. However, when the speed is not small or zero, then $(1 - v^2/c^2)$ is smaller than 1, and therefore γ is larger than 1. Thus the formula says that at some given speed, the momentum is larger than you might have thought. For example, at $v/c = 3/5$ (when the speed is three fifths of the speed of light) we have $v^2/c^2 = 9/25$, so $1 - v^2/c^2 = 16/25$, which makes γ equal to one over four fifths, $\gamma = 5/4 = 1.25$. As v becomes larger, so does γ, and, significantly, as v approaches c, equation (2.2) says that γ tends to one divided by zero, which is infinity. Therefore the momentum becomes infinitely large as the speed of the body approaches the speed of light—see figure 2.1. This tells us that the speed of light is special, and this fact will form a big part of our study of Relativity.

Using the gamma symbol to keep things succinct, Newton's Second Law takes the form

$$\text{force} = \text{rate of change of } (\gamma m\mathbf{v}) \tag{2.3}$$

Our first major lesson in Special Relativity is:

> If all you want is to be able to calculate how things will move, then we have now finished our study of Special Relativity!

You do not need anything other than this modification of the Second Law in order to calculate correctly all the motions of real physical bodies, moving at whatever speeds, as long as gravitation is not needed (you need General Relativity for that) and as long as quantum effects are negligible! In order to apply the equation in practice, one would need to know the size of the force in a given situation. For the case of a charged particle in an electric field, it is given by $\mathbf{f} = q\mathbf{E}$, where q is the charge of the particle and $\mathbf{E}$ is the electric field.

2.2.1 MOTION UNDER A CONSTANT FORCE

Let us immediately solve our first example problem: the motion of a particle under a constant force $\mathbf{f}$. That is, the force does not change with time, and does not depend on position. It could be produced, for example, by a uniform electric field acting on a charged particle. A uniform electric field can be obtained by putting an electric voltage between two large parallel metal plates: see Figure 2.2. It is easy to solve the relativistic equation of motion (2.3): we find that the relativistic momentum simply increases uniformly with time. That is, if a constant force acts on a particle for time t, then the increase in momentum of the particle is $\mathbf{f}t$. Therefore, if the particle started out with momentum $\mathbf{p}_0$, then later on its momentum is

$$\mathbf{p} = \mathbf{p}_0 + \mathbf{f}t \tag{2.4}$$

We will see later that we have to be careful in defining what we mean by time, and distance also, but for the moment let us just

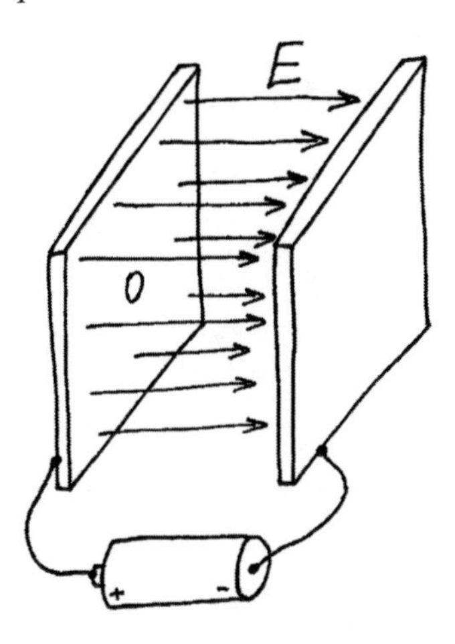

Figure 2.2: Creating a uniform electric field by applying a voltage to a pair of parallel copper plates. A *linear accelerator* consists of a sequence of such plates, with a hole in the middle to allow particles to pass through. Charged particles such as electrons are thus accelerated to high velocity.

note that this simple behaviour is what the equation of the Second Law predicts. Note, however, that in the Special Relativistic prediction, while the rate of change of momentum is constant, the acceleration is not.

Equation (2.1) shows what the momentum is if the mass and velocity are known. We are now in the situation where we know the momentum at any given time t, and would like to determine the velocity. This can be done by manipulating equation (2.1). In mathematics this is called 'inverting' the equation. The answer is

$$v = \frac{p}{\sqrt{m^2 + p^2/c^2}} \tag{2.5}$$

If you would like to see the proof of this, then consult the box—I encourage you at least to look through it. The essential idea is that if we had a relation such as 'p is twice v' then we could deduce that 'v is half of p'. Going from equation (2.1) to equation (2.5) is essentially a more complicated example of this basic idea. I put the proof in a separate box because it is not necessary to follow this particular proof in order to follow the rest of the book: we will not need it again.

Deriving the equation for velocity

The equation for momentum (2.1) can be converted into an equation for velocity as follows. First note that the momentum and velocity are always in the same direction. This means we already know the direction of the velocity: it is along $\mathbf{p}$. All that remains is to determine the size of the velocity: the speed v. In equation (2.1) there is a square-root sign, which makes things complicated. Therefore, we use the equation to devise a formula for $p \times p$, 'p squared':

$$p \times p = \frac{mv}{\sqrt{1 - v^2/c^2}} \times \frac{mv}{\sqrt{1 - v^2/c^2}} = \frac{m^2 v^2}{1 - v^2/c^2}$$

Next multiply both sides by $(1 - v^2/c^2)$, to obtain

$$p^2(1 - v^2/c^2) = m^2 v^2$$

Swapping the sides and multiplying out the bracket, this is

$$m^2 v^2 = p^2 - p^2 v^2/c^2$$

The last part on the right has v^2 in it, so now add $(p^2 v^2/c^2)$ to both sides of the equation. This cancels out the last part, so now the equation reads

$$m^2 v^2 + (p^2 v^2/c^2) = p^2$$

Because the order of multiplying does not matter (for example, 2×3 is the same as 3×2) this can be rearranged into $v^2 m^2 + v^2 p^2/c^2 = p^2$, and therefore

$$v^2(m^2 + p^2/c^2) = p^2$$

Now divide both sides by $(m^2 + p^2/c^2)$ and take the square root, obtaining

$$v = \frac{p}{\sqrt{m^2 + p^2/c^2}}$$

We now have the required formula that tells us the speed of a particle in terms of its momentum and rest mass, and the speed of light. This is equation (2.5).

Now we can see what happens to the velocity in our example case of motion under a constant force. Supposing the particle starts at rest (relative to the coordinate system we are using: more on this later), then in equation (2.4) we have $\mathbf{p}_0 = 0$, and so the equation says that $\mathbf{p}$ is always in the same direction as $\mathbf{f}$ and its size is $p = ft$. Putting this into (2.5), we find

$$v = \frac{ft}{\sqrt{m^2 + f^2t^2/c^2}} \tag{2.6}$$

This formula tells us the speed at any given time t, if the force f and mass m are known. To use the formula, we could, for example, consider a small object such as a plastic pellet of mass one tenth of a gram ($m = 0.1$ gram), subject to a force of size $f = 100$ newtons. (This force is about the same as the force you need to apply to pick up a 10-kilogram weight—moderately large, but not very large). After one second we would find that $ft = 100$ (newton-seconds), and therefore $ft/c = 3.3 \times 10^{-7}$ kilograms, (one third of a milligram), by using the fact that the speed of light c is about 300 million metres per second. This value of ft/c is very small compared to the mass m, so during the first second the pellet accelerates just as it would without worrying about Relativity. After one second it attains a speed which is high by everyday standards—about 1,000 kilometres per second—but which is still small compared to the speed of light.

If the force is applied for five minutes, then something new emerges. Five minutes is 300 seconds, so now $ft = 30,000$ newton-seconds and $ft/c \simeq 0.1$ gram. This is large enough to make a significant contribution to the part inside the square root symbol in the formula (2.6). As a result, the speed of the pellet is not as large as would have been expected according to Newton's understanding of acceleration under a constant force. The complete behaviour of the speed increase as time goes on is plotted in figure 2.3. At first the speed increases uniformly with time, but after about five minutes this uniform increase is not preserved, and the *acceleration* becomes smaller and smaller as the speed increases, in such a way that the speed never reaches the speed of light.

This sort of behaviour is observed in machines called 'linear accelerators' that are used to accelerate charged particles such as electrons and protons to speeds approaching the speed of light. Such particles can be accelerated very rapidly owing to their small

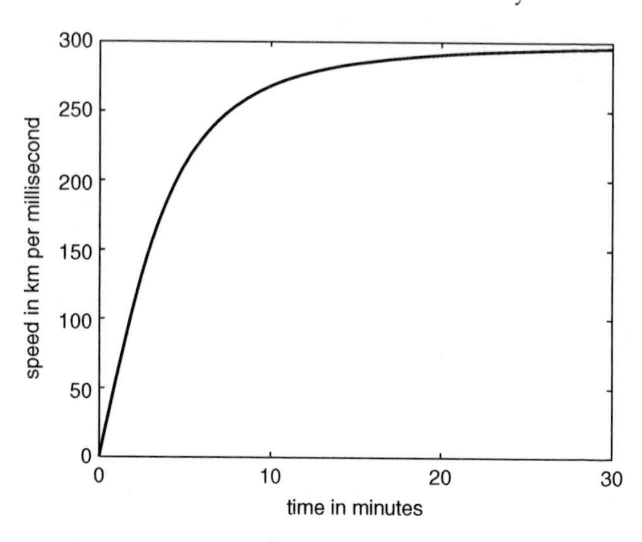

Figure 2.3: Motion under a constant force, for a particle starting from rest. The graph shows the speed of the particle as a function of time. It increases rapidly at first, then less and less rapidly, never quite reaching the speed of light. The relativistic momentum, however, continues to get larger and larger, without limit. The example given is for a 0.1-gram mass subject to a 100-newton force.

mass, so the time required to accelerate from a standing start to close to the speed of light is a fraction of a second, not minutes. However, precisely the same formulae apply, and it is found that if the electric fields used to accelerate the particles are constant, then the speed of the particle (electron or proton) increases with time exactly as shown in figure 2.3. This offers some confirmation that the equations we are using are correct!

Let us finish this chapter with a reminder of the argument so far. In this chapter I introduced some terminology, and then I simply told you a few facts about Special Relativity, without giving any proof. The main fact is that Newton's Laws of Motion are not changed, as long as we use a new relationship between momentum and velocity (equation 2.1), which leads to the 'Relativistic Second Law', equation (2.3). For illustration we then discovered what happens when a body is pushed by a constant force; for example, as would happen for a charged particle in a constant electric field. The main purpose was to show that we can make some useful progress without knowing much about Special Relativity. However, the real joy of the subject is not revealed simply by looking at the Laws of Motion. It is much more insightful, and more fun, to take another approach where we reason carefully about space and time and speed and relative motion. This is the subject of the next few chapters.

Part 1

Introducing space and time

3

Something odd is happening all around us

3.1 A shower of muons

Let me introduce you to a small particle that will play a large role in our learning about Special Relativity. This particle is the *muon*. It is a small, negatively-charged particle, very much like an electron except a lot heavier (about 207 times heavier). What is crucial about this little fellow, for our purposes, is that he does not live forever. Every muon breaks apart, after a short time, into three smaller particles: one electron and two other ones called neutrino and anti-neutrino. This breaking-up of the muon is caused by the fundamental behaviour of the forces which hold it together. There is no way to avoid it, and it happens according to a very strict timetable: for every microsecond that goes by, the muon has precisely a 63.43% chance of surviving. This means that the probability that a muon will survive for two microseconds is $0.6343 \times 0.6343 = 0.402$, for three microseconds is $0.6343 \times 0.6343 \times 0.6343 = 0.255$, and so on. After each period of 5 microseconds, the probability of survival falls by about a factor 10.

In short, if you start with one million muons sitting at rest on a table at some time, then after 10 microseconds there will be about ten thousand left, and after a total of 20 microseconds have passed there will be about one hundred muons left. Note that this is a tiny fraction of the number you started with: it is like saying almost none are left.

Now, if you possessed a muon detector (and some people do), you could observe something interesting: your detector will register a large number of muons entering it. At sea level, about 170 muons fall each second on each 1-metre-sized square area of the Earth's surface (about the size of a table-top). That is about one per minute per square centimetre (the size of a fingernail): something like a very light shower of rain, but going on all the time. These muons have high speeds, close to the speed of light, and in fact muons are the most numerous energetic charged particles found at sea level. Where have they come from? It turns out that most of these muons were produced by collisions in the upper atmosphere, at about 15 kilometres above sea level, when fast-moving particles (mostly protons) from outer space hit air molecules—see figure 3.1. The muons move nearly straight downwards (the effect of a slant angle to their trajectories would only make the results we are about to consider even more

Figure 3.1: The muon shower. When cosmic-ray particles hit the upper atmosphere, they generate showers of muons which move quickly towards the surface of the Earth.

surprising). By carrying a detector up a high mountain, or by putting one in a balloon, it is easy to measure the number of muons arriving at different heights. If you do this, a *very* remarkable thing is observed: **a large fraction (about a third) of the muons created in the upper atmosphere manage to travel right down to the Earth's surface!**

At first sight, this observation may not seem remarkable to you, but think a little. The muons are travelling at close to, but no faster than, the speed of light, which is 299, 792, 458 metres per second—about 0.3 kilometres per microsecond. Therefore, to travel the distance of 15 km from the upper atmosphere to the surface of the Earth, they must take $15/0.3 = 50$ microseconds. That is ten periods of five microseconds, so according to what we just said about muon survival, the probability that any muon could survive that long is $(1/10)^{10}$—0.0000000001, or one part in ten billion. Yet the muons come. Therefore, what is observed every day appears to be impossible.

Now, you should not just accept an impossibility. You should think about it some more. The observations suggest that the fast-moving muons can survive on average about 22 times longer than their normal lifetime of a few microseconds. So a natural explanation is just that: when moving fast, the muon becomes more stable and so lasts longer. Fine, but consider this: *how does the muon 'know' that it is moving?* What I mean is, suppose we think about the situation from the perspective of one of the muons travelling down through the Earth's atmosphere. As far as such a muon is concerned, it and all the other muons are not moving, but instead the Earth is moving towards them. So are we saying the movement of the Earth (relative to the muons) can affect their survival time? Surely this is too strange to contemplate! Maybe it is the Earth's atmosphere that is important. That does not help either, because the atmosphere is not very dense. It is mostly empty space with only a tiny fraction taken up by air molecules, and in any case collisions with air molecules have never been observed to increase the survival time of muons.

This muon survival observation is one of many pieces of evidence for Special Relativity, and in the next sections we will consider others, some of which were more important in the historical development of the subject. I have presented the muon example first because it will prove to be very useful later on. The main message you should remember is:

Main message: *The muons reach the Earth's surface, despite the fact that with their normal lifetime they would break up long before getting there.*

3.2 The light from the stars

The next remarkable observation is that the light from the stars always arrives at planet Earth travelling at the same speed (299, 792, 458 m/s, which is about 299, 792 km/s). This is surprising, because the Earth itself, and the stars too, have themselves some high speeds of motion. From astronomical observations we can estimate the distance from the Earth to the Sun quite accurately. It is 149 million kilometres on average, and the shape of the orbit is almost circular. Using the formula 'π times diameter' for the circumference of a circle, we find the path of the Earth's orbit is 936 million km long. We know it takes 365.25 days to go around, so the orbital speed is ($936/365.2 = 2.56$) million km per day, which is 107 thousand kilometres per hour, or about 30 km/s.

Now suppose we observe the light coming from a star that is located such that at some moment in one month, say January, Earth is moving directly towards it, while six months later, in July, Earth is moving away from it: see Figure 3.2. In January Earth is moving towards the oncoming light, so we might expect the light to enter a detector on Earth at the speed $(299,792 + 30) = 299,822$ km/s, and in July the Earth is moving away from the oncoming light, so we might expect the light to enter a detector on Earth at the speed $(299,792 - 30) = 299,762$ km/s. The difference between these two speeds is 60 km/s, so we might

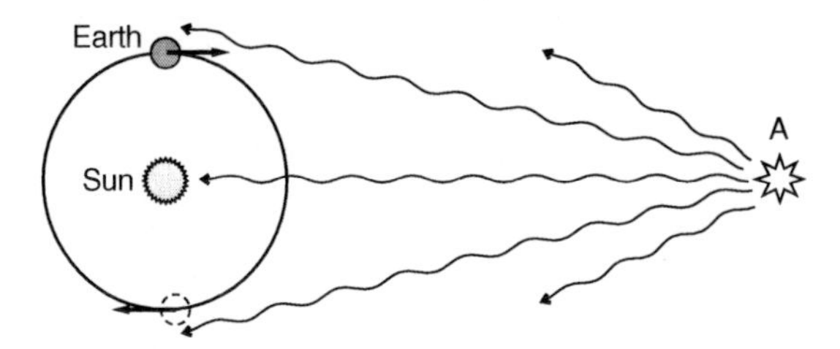

Figure 3.2: Owing to its motion around the Sun, sometimes the Earth is moving towards star A, and sometimes away. One might expect, therefore, that the speed of light from the star, relative to the Earth, would sometimes be higher, sometimes lower. However, it is found to be always the same.

expect the speed of starlight, relative to the Earth, to change by this much during the course of any given year. Although this is only a small fraction of the total speed of light, it is easy to observe it using good optical instruments . . . except that *no change at all is observed!*

To come clean, I should mention that the observations are not usually quite as direct as I have just implied. One cannot easily measure the speed of light directly—especially the dim light from stars—but one can deduce whether or not it is influenced by the motion of the Earth by combining various observations. One observation is that the angle at which the light from a star arrives on Earth varies during the course of the year. The part of this change in angle that is associated with the velocity of the Earth is called 'stellar aberration'. It is a small effect (about one hundredth of one degree) but measurable, and it therefore provides us with some useful information about the speed of light. On its own it is not surprising, but the challenge is to combine this observation with others we shall consider next.

To obtain more information, many ingenious experiments were carried out to examine light propagation in various circumstances. One of the most famous is the Michelson–Morley experiment of 1887: see Figure 3.3. This concerned the propagation of light between a set of mirrors fixed relative to the Earth. Of equal historical importance were studies on the propagation of light in moving water, carried out by Fizeau in 1851 and 1853. The result

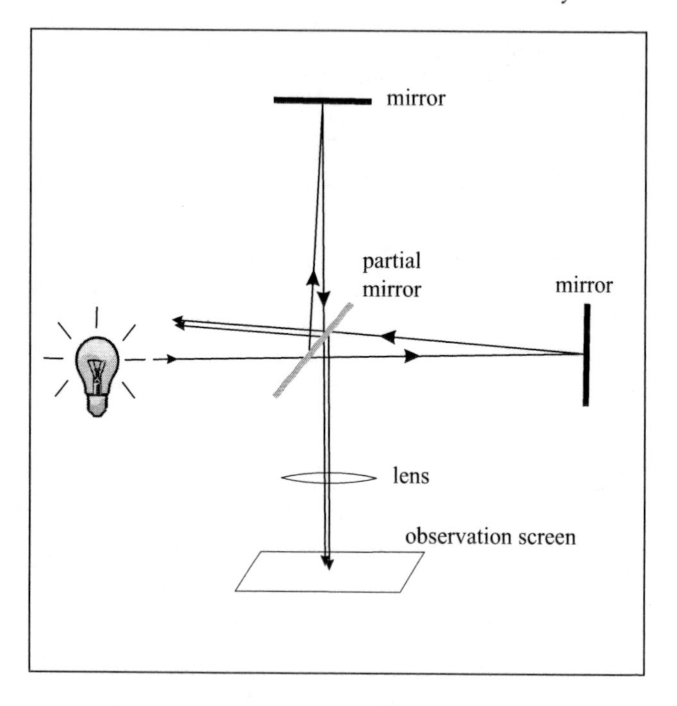

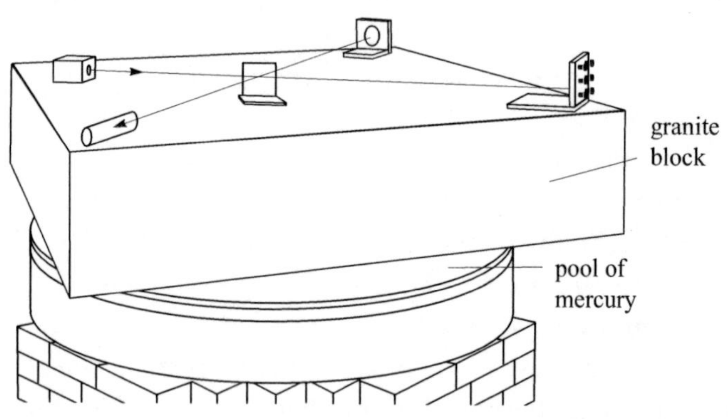

Figure 3.3: The Michelson–Morley experiment.

Michelson–Morley experiment

(see Figure 3.3)

A light source such as an ordinary light bulb is used to illuminate an optical device called an *interferometer*. The main components of the interferometer are a 'partial mirror' and two ordinary mirrors. The 'partial mirror' is one which imperfectly reflects the light: half passes through and half is reflected. It can be made, for example, by coating a flat piece of glass with a very thin layer of silver. The resulting reflector is called a 'beam splitter', because when the light hits it at an angle of 45 degrees, the result is two beams of light, propagating at right angles to each other. Two mirrors are positioned so as to reflect each of these beams back towards the beam splitter. The beam splitter once again reflects or transmits the light hitting it. Because light is wave-like, the returning light waves from the two paths either reinforce one another, or cancel one another out, depending on whether or not the two sets of waves are in step (this is called 'interference'). This is useful because it is a very precise effect, easily able to reveal distance changes in the interferometer of just half a wavelength of visible light (about 0.25 thousandths of a millimetre) while the distance of travel is some metres, so a sensitivity of one part in ten million is available. This is also the sensitivity to changes in the travel time of the light in the two parts of the interferometer. Since the whole interferometer is travelling along with planet Earth, one might expect the speed of light along the two paths to be different, and by rotating the interferometer through 90 degrees it is easy to check for this. However, no difference was found. For a further check, the experiment was carried out repeatedly in different months of the year. Michelson and Morley first performed their experiment in 1887, and found to their surprise that they could rule out a change in the speed of light at the level of one quarter of the Earth's orbital speed. To reach to this precision they needed a set of eight pairs of mirrors mounted on a granite slab floating in a pool of mercury. They and others repeated the experiment many times in the period 1887 to 1930, at greater levels of precision. (An earlier experiment by Michelson in 1881 was inconclusive; his collaboration with Morley involved several ingenious improvements to the apparatus to make it more precise. The diagram here has been slightly simplified.).

of all these studies is that although light travels slower in a medium such as glass or water, its speed in vacuum is never in the least altered by motions of either the light source or the measuring device.

Just as with the muons, you should not read about an observation like this without trying to think of a way to explain it. Many extremely able scientists thought hard about this over a period of fifty years in the second half of the nineteenth century. One reasonably natural explanation appeals to the wave-like nature of light. Most waves, such as waves on water, or sound waves in air, require some medium in which to propagate (for example, water or air). Think of waves on water to have a definite example (you could try experimenting in a bath). Suppose water ripples travel across the surface of the water at one metre per second. If there is a source of ripples, and you move (in a boat, say) towards the source at one metre per second, then as the ripples pass your boat you will see that their speed, relative to your boat, is two metres per second. If you travelled the other way, you could keep pace with the ripples and then their speed relative to the boat would be zero. However, think about what happens very close to the boat. The water there does not slip past the boat completely freely. Instead, the water clings slightly to the sides of the boat, with the result that the boat drags a thin layer of water along with it. If you could measure the ripples *in this thin layer*, then you would see their speed, relative to the boat, is always one metre per second, because the layer is not moving relative to the boat.

So, the explanation for the constant speed of light ran like this: although light is observed to be able to propagate in vacuum, it must be that the 'vacuum' is not really empty, but contains some sort of medium, and light waves are an oscillation of this medium. The medium was called 'aether'. The lack of any change in the speed of light as the Earth travels along its orbit was attributed to a dragging of some substantial volume of the 'aether' around the Earth, just like the dragging of the layer of water close to the boat.

However, this 'aether dragging' hypothesis does not account for all the observations. It could be used to account for the Michelson–Morley result, but not the stellar aberration (unless it is 'fixed up' by further adjustments to the argument), and it can only be made to fit Fizeau's results if a moving medium such as water only partially drags the aether. In short, there were a lot of problems with this aether theory. It took careful reasoning to apply it to each experiment, and as more and more different types of experiment were tried, more and more convoluted and strange properties had to be attributed to the aether. It began to look like the aether idea was simply wrong and explained nothing.

The long and short of it is, light behaves in a very startling way.

3.3 The Maxwell equations

It would not be right to finish a chapter presenting some of the important evidence for Special Relativity without mentioning the Maxwell equations. This is a set of equations developed by various physicists and completed by James Clerk Maxwell in 1861, which brought together all that was known about electricity and magnetism, and furthermore predicted the possibility of electromagnetic waves. In Maxwell's own words: 'We can scarcely avoid the conclusion that light consists in the transverse undulations of the same medium which is the cause of electric and magnetic phenomena.'

The Maxwell equations are one of the greatest achievements of physics. Their mathematical structure turned out to be a forerunner of Special Relativity. They are the first relativistically correct set of equations to have been discovered, and they underpin most of the great discoveries of physics since the beginning of the twentieth century, including Relativity and quantum field theory. For this and his many other contributions, James Clerk Maxwell should have, it could be argued, a place in our general esteem close to that of Newton and Einstein. The city of Edinburgh can, and I think does, pride itself on being the place of his birth and early education.

Maxwell's equations were used by several theoretical physicists, especially Hendrik Lorentz (1853–1928), to tease out aspects of Relativity, before Einstein clarified the whole subject. For example, they predict correctly that the speed of light in vacuum is independent of the motion of the source. As Einstein showed, this implies some revolutionary ideas about time, and rather than accept this, attempts were made to construct modified theories that could encompass the experimental observations, while remaining consistent with a traditional notion of time. The most notable is the 'emission theory' (1908–11) due to Ritz. This could explain the experimental evidence available in 1911, but it was soon ruled out by observations of binary stars and by Michelson–Morley experiments carried out using the Sun or another star as light source. Maxwell's equations survive to this day. Together with quantum theory they underpin all we know about light and the way by which charged particles interact with one another.

3.4 Conclusion

In this chapter we have considered two types of experimental observation: the long-lived muons, and the constant speed of light. The main message I want to convey is that the observations that we are dealing with concern quite basic aspects of the world: how long something lives, and how fast something moves. This involves distance, time and speed. I could present many more examples, but all I want to do for now is to leave you with the feeling that something odd is happening all around us, and the strangeness is quite pronounced. The technical details of the Michelson–Morley experiment might hide the main message, so let me repeat it:

> If you set up a light bulb, and stand still relative to it, then the light will zoom past your body at close to[1] 299, 792, 458 m/s. If you run towards the light bulb, or make it move towards you, the light will still zoom past your body at 299, 792, 458 m/s. Even if you put the light bulb on a rocket (11, 000 m/s) or on the planet

[1] Here I am ignoring the refractive index of air.

> mercury (48, 000 m/s) or on a super-duper rocket (moving at, say, 200, 000, 000 m/s) there would still not be the least change in the speed at which the light moved past you. Equally, if you try to chase the light moving away from you, by jumping aboard ever faster rockets, you will never catch up with it, or even alter by one iota the speed at which it recedes from you.[2] Evidence from astronomy shows, furthermore, that this is not special behaviour found only on Earth: it is true throughout the solar system, and throughout the galaxy. *It seems to be a basic property of the universe.*

The constancy of the speed of light, and the muon experiment, require an explanation, which will be provided in Part II. The next chapters will prepare the way by introducing the concepts we will need.

[2] This nice phrase is from Wolfgang Rindler.

4

Spacetime

In this chapter we will study only 'traditional' physics, which does not need Special Relativity, but we will do so using a method that will be very helpful when it comes to learning Relativity. We will learn about 'spacetime' and spacetime diagrams. 'Spacetime' means just what the word suggests: space and time, and it is no great surprise that we might want to think about an idea like that. However, in pre-Relativity physics one did not have to think about spacetime as a single idea: one could always think about space and time separately—space providing the 'arena', as it were, for where things happen, and the passage of time being a way of allowing for the fact that things change. This clear separation will no longer be valid when we come to learning Relativity, where space and time are more intimately intertwined. However, it is a very helpful aid to learning if we first accustom ourselves to the idea of spacetime while discussing familiar things, such as the motions of ordinary objects at low speeds.

To repeat:

> In this and the next chapter there will be NO use of Special Relativity. EVERYTHING will be in terms of the laws of physics as understood before the twentieth century. These are called the *Classical laws of physics*. They are accurate to approximately one part per billion when all relative speeds are below 10,000 m/s, gravitational fields are weak, and quantum effects are negligible.

4.1 Time

We all have a fairly good intuition about space, about distances, and so on, but it is famously hard to have quite such a good intuition about time. Time is ultimately mysterious, and the study of Relativity will not change that. However, what Einstein himself emphasized, and what his theory leads us to take seriously, is the idea that we do not need to have an abstract philosophical definition of time. In fact, the attempt to formulate one might turn out to be misleading. Rather, instead of talking about time in the abstract, we will talk about clocks: that is, physical objects which change in some regular way. Ultimately all we are going to attempt to do is provide some sort of reasonable and satisfying framework to describe the way physical objects behave, including objects such as mechanical ones with gears and springs (a bicycle, an alarm clock) or electrical ones (a radio, a computer, a digital watch) or oscillating atoms, or radioactive power-stations, and so on.

Suppose that some regularly repeating object such as a well-made watch is made to emit ticking noises. Suppose that other objects are near to the first one, and we arrange that no external forces act (which could affect some objects differently to others). If for each tick, all the other objects go through a regular number of their oscillations, or a standard sequence of their internal dynamics, then we see that they all have something in common, and for brevity we call that something 'time'. This is a very useful word, but if at any moment we are unsure what we mean by it, then go back to thinking about the clocks—the physical objects with a regular behaviour.

4.2 Spacetime diagrams

We are now going to discuss the simplest type of motion: motion in one dimension at low speeds. In order to obtain a really clear understanding, I would like to encourage you to construct a simple aid to learning. Take an ordinary wire, reasonably stiff (for example, cut it from a metal coat-hanger), and about 30 cm long, and get some beads or small cylinders that can slide along it. You could use, for example, cylindrical pieces of pasta, or cut some lengths off a plastic pen, some a few millimetres long and some 2 centimetres long. It will be useful if you can have some of the beads or sliders with a hole large enough that another can slide right through it. Put your wire and beads conveniently close to hand as you read.

We are going to discuss beads sliding along such a wire. If you cannot get hold of a suitable wire, then use another visual aid, such as the edge of a table, and just imagine things sliding along it. The main thing is to be clear that the objects we shall discuss will only move forward and back along a line; they will not veer off to one side. Other examples of motion in one dimension are trains running along a straight track, or spaceships moving in a straight line, far from any planet or star.

First place a single bead on your wire, motionless in one place. We can represent the story of this bead by a simple graph. Let the *horizontal* axis represent distance along the wire, and the *vertical* axis represent time (it is important to have the axes the right way round). Then the 'life history' of the bead is represented by a vertical line. This shows that the position of the bead, as time goes on, is unchanging: see Figure 4.1a. Note that you may be used to plotting motions 'as a function of time', where one puts the time axis horizontally. There is nothing logically wrong with doing that, but it will turn out to be clearer to think about spacetime with the time axis vertical on the diagram. Try to get used to this way of doing things.

Now take another bead, and slide it gently along the wire with your finger, keeping the speed constant, until it hits the first bead and stops. This simple experiment is represented by the graph shown in figure 4.1b. We see the vertical line representing the first bead, staying still, and a sloping line representing the fact that the second bead moves along as time goes on. In my graph I have indicated an example of motion from left to right. I could have

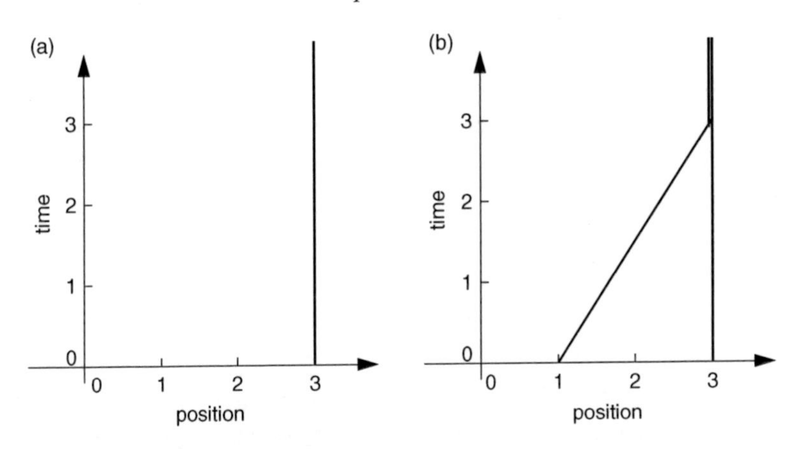

Figure 4.1: Diagram showing the position of beads on a wire, as time goes on. In the left diagram (a), a single bead stays at a fixed position. In the right diagram (b), there are two beads. One stays at a fixed position, and the other moves towards it and then stops.

chosen the other direction, in which case the line would slope the other way.

The graph we have drawn is called a *spacetime diagram*. It summarizes all the behaviour of the objects (here, the beads) in space and time. To acquire a good sense of this, look at the spacetime diagram another way. Cover almost all of it up, using either two pieces of paper with a small horizontal gap between them, or by using a single piece of card with a long narrow horizontal slot cut in it. Through the slot, or in the gap, you can now see one 'slice' out of the graph, and it shows the situations of the two beads at some given instant of time. By sliding the slot up the diagram, you can follow the behaviour of the beads as a function of time, like watching a movie. You will see one 'bead' (really a line of ink representing a bead) slide along until it hits the other. I would like to encourage you to actually do this: make the wire, slide the beads on it, make the slot, and slide it up the diagram. Do this several times until you are very happy with the way the diagram can be used to show the motions of the beads. If having done all this you are not smiling, then perhaps you had better put down this book, because it is for people who think that things like this are fun.

A sufficiently small bead would be what we call in physics a 'particle'; that is, an object small enough that to all intents and purposes it has no size. Notice that such a 'zero-dimensional' object appears on a spacetime diagram as a *line*, which is a one-dimensional thing. This is because the object of negligible *spatial* extent still has a large 'extent' through *time*. The line on a spacetime diagram that represents the history of a particle is called a *worldline*. You should not think of a particle as 'moving along' its worldline. Rather, the worldline *is* the particle. A particle has an extended existence in time, and therefore from a spacetime point of view it is a one-'dimensional' object, though the 'dimension' extends through time not space.

We have just thought about something quite interesting. This business of a particle *being* a worldline, not 'moving along' one, is worth mulling over: perhaps you might like to sleep on it.

Now let us proceed to an object of larger size. An object having one spatial dimension is something like a long thin cylinder or a piece of thin wire. We can indicate the behaviour of an object like this on a spacetime diagram by marking in the worldlines of the two ends of the object. Do the following experiment: get a narrow cylinder and slide it along the wire, and pass it right through a fixed bead (you will need one with a hole that is large enough). The spacetime diagram for this experiment is shown in Figure 4.2. Notice that the object having one spatial dimension looks like a two-dimensional region in spacetime: the region is shaded on the diagram.

Again use the 'slot' method to follow the motion, and thus watch on your 'movie' as the cylinder approaches the bead, then for a while the bead is within or over cylinder, then it goes out the other side.

Figure 4.3 shows a more complicated example: the motion of the balls in 'Newton's cradle'—a small toy illustrating the physics of collisions. You should be able to relate the diagram to the swinging motion of the balls. Do not forget that the diagram shows

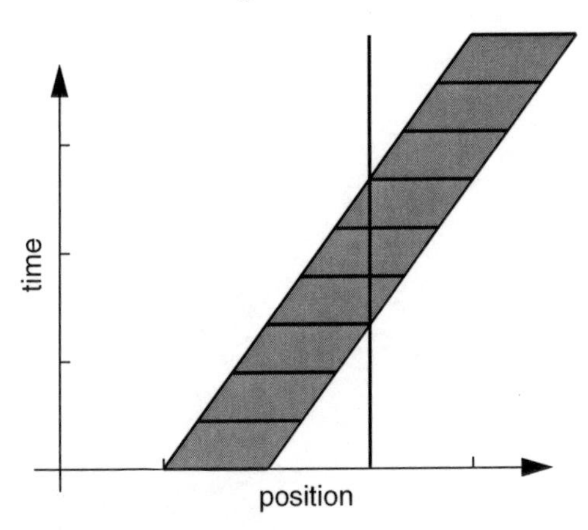

Figure 4.2: Spacetime diagram showing a bead and a one-dimensional object such as a thin stick or rod, both evolving through time. The rod moves towards and then passes over the bead. Note this diagram does not show a 'sloping rod', it shows a rod that is at all times lying along the one-dimensional wire. The bars show the position of the rod at a selection of times. The shaded region is the region of spacetime occupied by all the material of the rod, as it evolves through time.

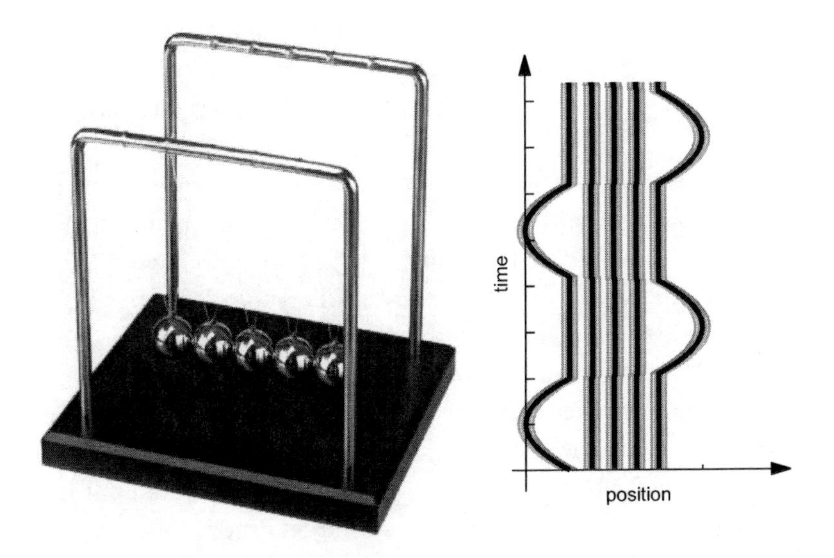

Figure 4.3: 'Newton's cradle'. On the left is a picture of the toy called 'Newton's cradle', consisting of balls hanging in a row, so that they can swing and bump into one another. On the right is a spacetime diagram showing the horizontal motion of the five balls, when the ball at the left is first moved to one side and then released. There is also a small amount of vertical motion, which is not shown here.

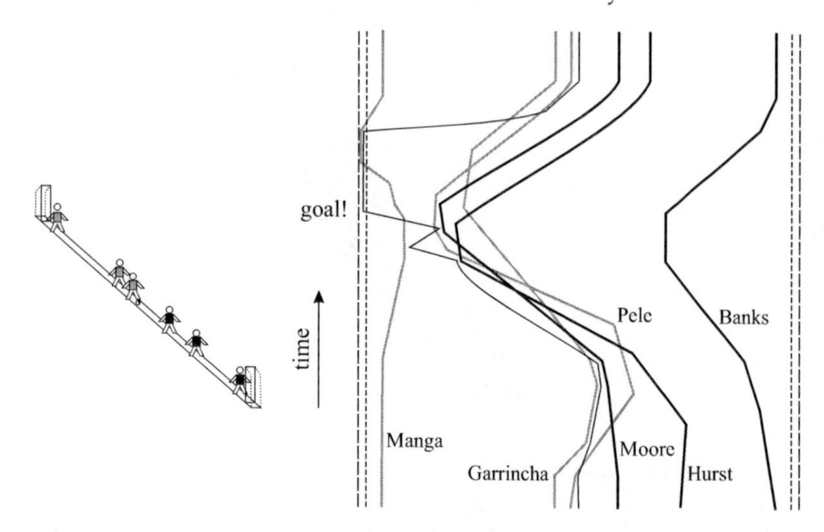

Figure 4.4: Linear football. 'Pelé nudges it and runs forward; picked up by Garrincha; nice tackle by Moore; he kicks it up the field; Hurst runs through; he's got a chance, he shoots! Saved by Manga!—but Moore gets the rebound; he strikes a volley—right through Pelé—goal!! Manga can't believe it.'

motion along a line as a function of time, and by looking at it through a horizontal slot you can remind yourself of this. Figure 4.4 shows a game of 'linear football' which is played on a long thin pitch.

So far we have talked about worldlines and two-dimensional regions, but our diagrams also have zero-dimensional points on them. An example is the point where the lines meet in Figure 4.1b, and each of the two points in Figure 4.2 where the line meets a side of the shaded region. A point in spacetime is called an *event*. To have an example of an event, click your fingers once, or if you cannot do that then tap your finger once on your forehead. That click or tap, brief and now over, was an *event*. It happened at one particular location, at one particular time, but it did not have any significant spatial extent nor duration in time.

The terminology we have introduced so far is summarized in the following table.

Puzzle. The beginning of what famous journey of 1969 is shown in the spacetime diagram in Figure 4.5?

Spacetime diagram	=	diagrammatic representation of the behaviour of objects in space and time, like a graph, usually oriented so that time goes upwards
Event	=	a point in spacetime
Worldline	=	line representing the history of a particle
Particle	=	physical object of negligible spatial extent

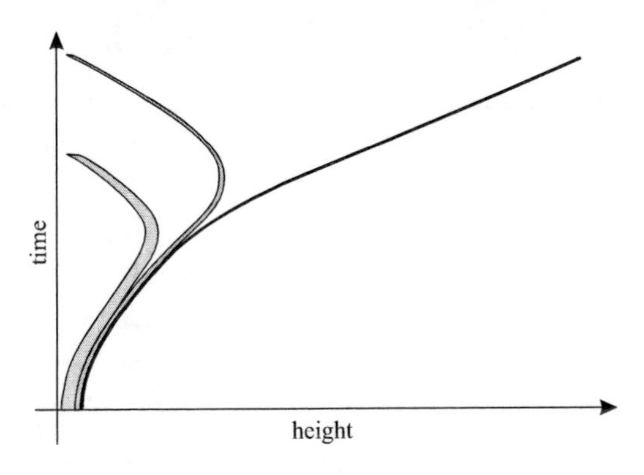

Figure 4.5: The start of what famous journey is shown (approximately) in this spacetime diagram?

Challenge. Sketch a spacetime diagram for a relay race. For simplicity, just show one team of runners, all running along a straight track. Each runner waits for the previous one to arrive, starts to run as they approach, then continues the race while the previous one slows to a stop.

4.2.1 MORE DIMENSIONS

In the above we considered motion purely in one spatial dimension, such as the motion of beads along a wire, or trains along a straight railway track. Motion in two spatial dimensions, such as the motions of billiard balls on a table, or of skaters on an ice rink, can be represented by a spacetime diagram with two spatial axes

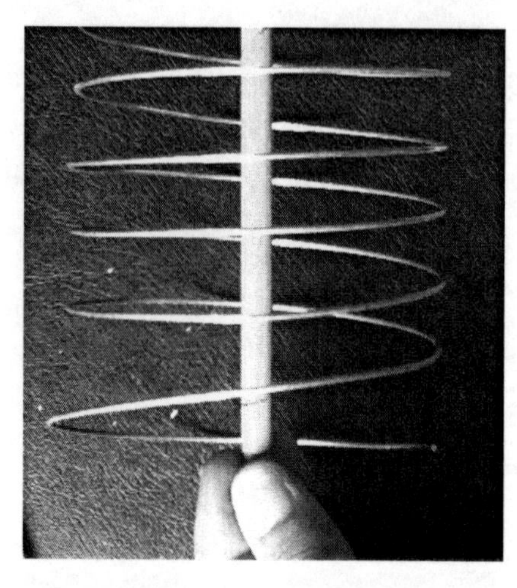

Figure 4.6: A photograph of a three-dimensional spacetime diagram. The diagram shows the worldlines of the Earth and the Sun, with one spatial dimension suppressed so that the time dimension can be shown.

and one time axis, making a three-dimensional diagram. Figure 4.6 shows a photograph of such a diagram (constructed using metal wires in air rather than ink lines on paper). The diagram represents the orbit of the Earth around the Sun. This orbit lies completely in a plane in space, to very good approximation, so a three-dimensional spacetime diagram can show it. The spatial situation at any given instant of time is obtained by taking a 'time slice' of the diagram; that is, by examining a thin horizontal cross-section.

The complete story of a game of snooker or billiards could also be shown using a three-dimensional spacetime diagram. Some other games, such as football or tennis, involve players that move in two dimensions and a ball that moves in three dimensions. The complete story of the movement of the players could be shown on a three-dimensional spacetime diagram, which could also indicate the horizontal motion of the ball.

Motion in three spatial dimensions would require a four-dimensional spacetime diagram. We cannot construct this directly in our three-dimensional world, but we can calculate what it would be like, and we can illustrate aspects of it by taking three-dimensional or lower-dimensional 'cross-sections' of it.

Puzzle. Draw a circle on a piece of paper. In spacetime, what is the shape formed by the set of worldlines of the ink molecules making up your circle?

4.2.2 SUMMARY SO FAR

If you have understood this discussion of spacetime diagrams correctly, it should begin to make you feel a little bit fatalistic. When we draw the behaviour of particles on such a diagram, we are implying that the future is already 'mapped out', and particles do not so much 'move about as time goes on', but rather simply 'exist as a line in spacetime'. If you and I have a worldline, is it too already 'mapped out', including all our eventual meetings and decisions? It is not clear whether the spacetime picture completely undermines the notion of free will, because if you take it seriously enough then it is not that your worldline will 'force you' to do anything: rather, you are that worldline, and it is you. Maybe the reason it twists one way rather than another in your future is because you will freely make one choice rather than another. In short, we do not understand free will, and it remains a matter of personal opinion whether or not we have it and what it means.

Some scientists claim that science itself implies that free will is an illusion, but others have pointed out that if this were so then the very process of reasoning on which the scientific method is based is undermined, because it would mean our thoughts are channelled and not able to be steered by an appeal to reason. This means the claim that there is no free will is highly dubious.

It is certain that our scientific understanding is incomplete. Although we have a good grasp of much of basic physics, it seems to me likely that the question of free will involves a new and subtle type of physical law that we have not yet formulated. The best approach to learning Relativity is to get used to the spacetime diagram and the spacetime way of thinking, and postpone to the end whether or not you think it impacts on determinism and free will. Certainly one has to be quite careful when applying spacetime reasoning to quantum physics, where some new subtleties come into play. For what it is worth, my own tentative conclusion (based on the evidence of human experience in general, not just physics) is that it makes sense to believe we have free will and all the creativity and accountability that accompanies it.

4.3 Tick tock

To introduce some more information into the spacetime diagram, let us think next about time. To be specific, as I have already mentioned above, we will think about physical objects that evolve in a regular way, such as clocks. Suppose you could put a small clock on top of one of the beads on your wire. If you have a wrist-watch, you could try it. It any case it is easy to imagine some sort of regular process happening at the location of a bead. The very molecules of which the bead is made can act as tiny 'clocks', because they rotate and vibrate in regular ways. The end of each repetition of any such process is an event. These events all happen at the bead. The sequence of these events or 'clock ticks' can

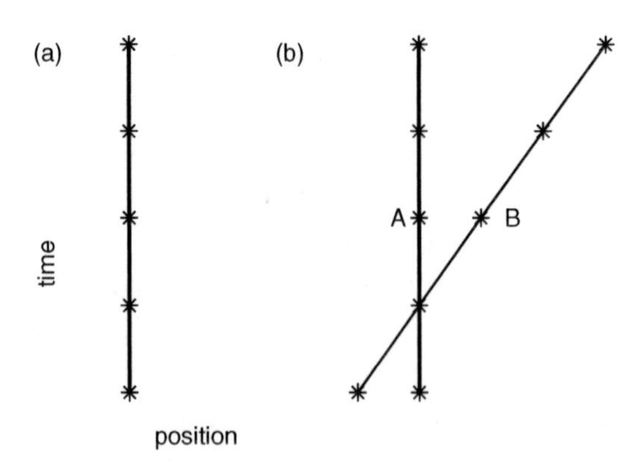

Figure 4.7: Adding timing information to spacetime diagrams. On the left diagram (a) a single bead carries a clock (some sort of regular repetitive process), and the successive tick events of this clock are indicated by marks on the world-line of the bead. On the right (b) two beads are shown, each carrying the same type of clock. Events *A* and *B* each occur at one clock tick after the meeting of the beads, so they are simultaneous on this classical spacetime diagram.

be marked on the spacetime diagram as a sequence of regularly spaced points on the worldline of the bead: see Figure 4.7a.

What about the case of a moving bead? We imagine what would happen if a clock were attached to the bead. By a 'clock', again we mean not some abstract entity, but a physical thing—for example, made of gears and springs—or it could be just one of the molecules of the bead. According to classical physics, as long as a system of objects is not rotating as a whole, and is not subject to external forces, its behaviour is the same as if it were not moving, except that the whole system moves along at constant velocity. Indeed, this principle is absolutely central to Relativity and we will return to it again and again. We apply it to a 'system' consisting of bead plus clock. It implies that a clock attached to a moving bead will tick away in just the same way as one attached to a stationary bead, as long as the velocity of the bead is constant. Therefore, we can mark the clock ticks again as equally-separated events, now on the worldline of the moving bead. Before we do so, however, we have to consider whether these ticks should be separated by the same amount as those that we have already added to the worldline of the first bead.

First of all, let us assume we are using the same type of clock for the two beads. For example, the ticks in both cases could be at the rate of one per second. Also, when the two worldlines cross we can arrange that the two clocks agree, so we have a useful 'tick event' in common. Before and after this, however, the clocks are separated. How can we ensure that the diagram indicates their behaviour correctly? Let us assign the name 'event A' to the event when the stationary clock indicates '1 second since the crossing' (it is marked A on Figure 4.7b) and 'event B' to the event when the moving clock indicates '1 second since the crossing'. According to classical physics, time 'marches on' equally for all things everywhere, so events A and B are simultaneous. This means that on our classical spacetime diagram, events A and B must be at the same vertical height on the diagram. Similar reasoning applies to all the other clock ticks, and we end up with a diagram as in figure 4.7b.

You should now be able to see that the passage of time, in a classical spacetime diagram, can be envisaged as a series of horizontal lines. Each horizontal line is the set of all events that are simultaneous with any given event on the line. We say that they 'occur at the same time as' the given event. We already implicitly assumed this when I encouraged you to look at the diagram through a horizontal slot, but by thinking about physical events such as the oscillation of a molecule or the ticking of a watch we have spelled out what we mean by the passage of time. Such events also allow the diagram to speak for itself. This is why in figure 4.7b I did not label the axes. After all, spacetime does not come with axes attached to it: it is just itself.

4.3.1 RELATIVE MOTION

A keen reader will have noticed that I slipped into some slightly lazy use of language in the two previous sections. I talked about 'the stationary bead' and 'the moving bead' without saying relative to what they were still or moving. I had in mind stationary or moving with respect to the wire. However, in studying Relativity it is crucial to try to move away from the notion of a 'fixed background' such as the wire in our example. We want to focus our minds on *relative motion*. Try to imagine, if you can, what it would be like if the beads moved to and fro along a line just as before, but there is no wire: they are just floating in space. For clarity, put them in outer space where there are not even any planets nor stars nearby. Now it becomes clear that there is no way to say which bead is 'at rest' and which is 'moving': all we can say is that there is relative motion between the beads.

What this means about the spacetime diagram is that we have to get away from giving either of the worldlines on our diagram any special status. The worldline that appears vertical was the one we called 'stationary' up till now, but of course either bead can be regarded as 'stationary': it depends which one you choose to be the one you are 'sitting on'. Or if you have some other motion, different to that of either bead, then both beads are moving relative to you.

As a first step to disposing of any special status for one bead rather than another, let us distinguish them another way. Suppose both beads are small, but one is noticeably bigger than the other. We will call them 'the thick bead' and 'the thin bead': then it is easy to indicate them on a spacetime diagram using a thick line and a thin line respectively. In fact, this has already been done in Figure 4.7b in readiness for the next part of the discussion.

Now we will draw the spacetime diagram again, but instead of making the worldline of the thick bead vertical on the diagram, we will choose some other arbitrary slope. In order to represent the same *relative velocity* of the two beads as before, the worldline of the thin bead will also have to change slope. Its slope must be chosen in such a way that the distance between the two worldlines, as measured along the distance axis (which is horizontal on the diagram), increases at the same rate as before, per clock tick. We thus obtain the spacetime diagram shown in Figure 4.8. It is important to note that this spacetime diagram is just as valid as the one we drew before (figure 4.7b). Both diagrams are correct representations of the *same* set of events in spacetime. It should not concern us that some aspects of the diagram have changed. It is not unusual for diagrams that look different at first sight to be showing the same thing really: Figure 4.9 shows an example concerning a cubic wire frame. You should be able to see that although Figures 4.9a and 4.9b are different in some respects, they show the same essential properties of a cube, such as the number of faces and edges, the edges are straight lines, three meet at each corner, and so on. They both show the same cube. In a similar way, Figure 4.7b and Figure 4.8 both show the same region of spacetime.

To make the point more strongly, I have shown another example in Figure 4.10. There are three versions, but all show the same region of spacetime. The events shown are as follows: at the bottom left and bottom right are two beads that have initial velocities away from each other, but after a while they stop and then move

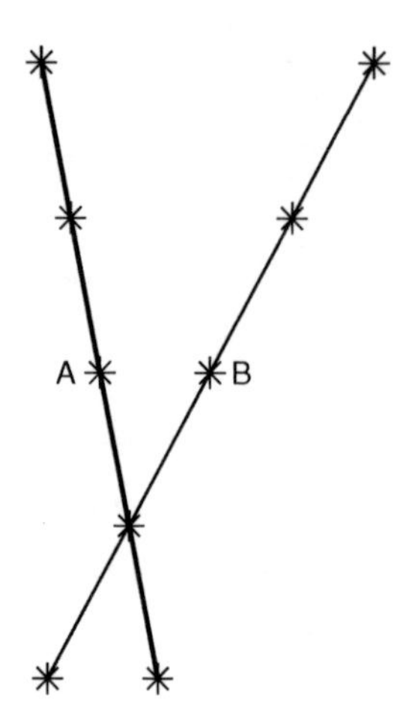

Figure 4.8: Spacetime diagram showing precisely the same events and worldlines as Figure 4.7b.

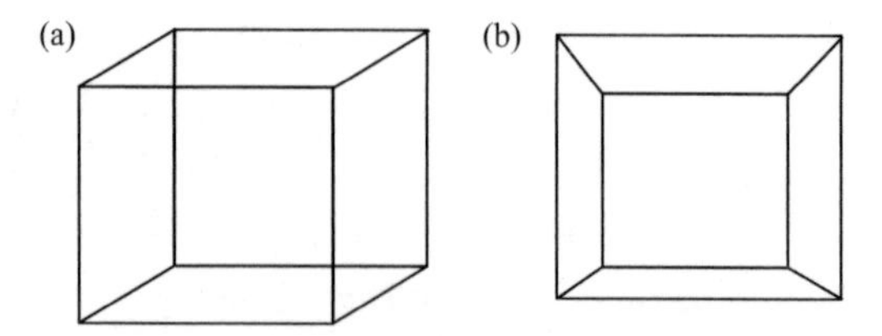

Figure 4.9: Two diagrams of a cube. Although the diagrams look different, they both show the same object: a cubic wire frame. The diagrams do not show two different cubes, but one cube which is being pictured in two ways. This illustrates the fact that diagrams in general can undergo changes which do not necessarily signify any change in the thing being examined; only a change in how it is being presented. Therefore, it should not unduly trouble us that the same applies to spacetime diagrams.

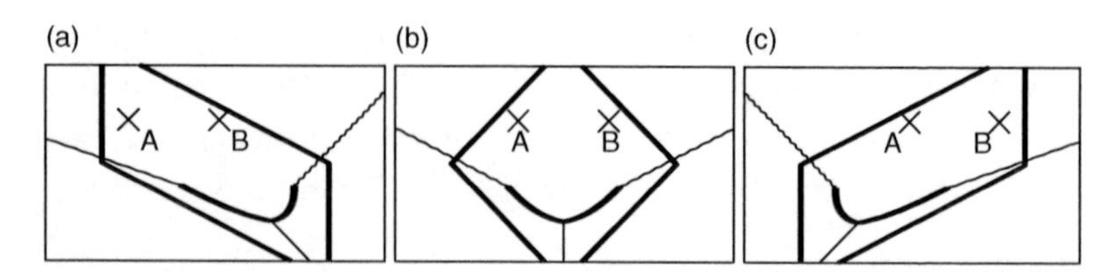

Figure 4.10: Three diagrams showing the same region of spacetime. The set of events is as described in the text: a pair of beads initially move apart but then move back towards one another, while between them another bead has constant velocity at first, then splits into two fragments which decelerate and then change into fast oscillating particles. Two further events A and B are also marked. These three spacetime diagrams have different visual appearance, but they indicate precisely the same set of events: all times (vertical direction on the diagram) and relative distances (horizontal direction on the diagram) are the same.

back towards each other. While this is happening, another bead half-way between them does nothing at first, but then splits in two. The two fragments move away from each other, towards the outer beads, slowing a little as they go, and then changing into faster-moving particles which oscillate as they move. Two other unrelated events are also shown, marked A and B.

All the distance (horizontal) and time (vertical) relationships between events and worldlines are precisely the same in the three diagrams of figure 4.10a, b, c. This is what we mean when we say that they all represent the same region of spacetime. The fact that they look like skewed versions of each other is simply a limitation of the way diagrams on paper (completely in *space*) can represent events in *spacetime*. It comes from the fact that we are using a spatial distance on the paper to try to represent time, but although time and space share some aspects of their nature, they are not exactly the same.

There is an infinite number of ways to draw the spacetime diagram for any given region of spacetime, depending on our choice of how to slant the worldline of the first constant-velocity particle we select. Once this is done, the other events and worldlines can be placed relative to the first. The possible diagrams for a given spacetime region are all related by a simple action: to go from one diagram to another, you simply 'skew' the diagram, in the same way that a stack of cards, resting horizontally on a table, can be slanted one way or another by pushing against the side of the stack with a sloping straight edge. This is for the 'classical' description of spacetime, the subject of this chapter. When we come to Special Relativity in Chapter 8, we will find that spacetime diagrams undergo another sort of 'stretch'.

5

Reference frames and coordinates

When I introduced the spacetime diagram, I referred to it as a kind of 'graph'—and it is that. It is a graph on which we plot the position of various objects as a function of time (with time running vertically on the graph). However, next we will think about the fact that there is more than one way to assign coordinates to events in spacetime.

We will approach this issue first by considering some informal examples from everyday life, and then by constructing the right concepts and mathematical analysis more carefully.

5.1 Informal introduction

Consider the following example. In the game of golf, an important skill is to use a 'putter' to hit a golf ball (diameter 45 mm, mass 45 grams) so that it rolls along the ground over a distance of typically a few metres, in order to make it fall into a hole of diameter 108 mm. Usually this is done on a golf course, which is a pleasant park with grass and trees. The longest conventional putt, where we measure the length in the ordinary way, is possibly one made by Fergus Muir at St. Andrews gold course in 2003. He used his putter from the tee and obtained a hole-in-one over a distance of 375 feet (114.3 metres). However, in 1997 a keen golf-player performed this feat in the unusual surroundings of a passenger jet airliner. While the aircraft was in flight, the golfer putted the ball up the aisle. It travelled a distance of a few metres in the aircraft, taking a time of approximately 1 second, and entered a small plastic container serving as the 'hole'.

If the golfer wanted to assign coordinates to the events in this simple experiment, he could do so as follows. Let event A be 'the ball is first struck', and let event B be 'the ball reaches the hole'. We can create a coordinate system so that the origin is at the location of event A, and we can measure time intervals from that event. Then the time and spatial coordinate of event A are ($t_A = 0,\ x_A = 0$), and the time and spatial coordinate of event B are ($t_B = 1$ second, $x_B = 3$ metres).

Now, I hope this example experiment has already set you thinking about the following issue. If the aircraft is in flight, then we expect that the golf ball could have travelled a good deal further than 3 metres, relative to something other than the aircraft. Indeed, a passenger jet can easily attain a groundspeed (speed relative to the ground) of 500 miles per hour, which is 222 metres per second. So let us suppose the groundspeed of the aircraft is 222 m/s. Assuming the golf ball was hit from the back towards the front of the aircraft, in its 1-second journey it travelled $3 + 222 = 225$ metres relative to the ground. An observer standing on the ground would naturally choose a coordinate system fixed relative to the ground. Let us use t' and x' for time and position in this new coordinate system. The ground observer can agree to measure time and distances from event A, so he would set ($t'_A = 0,\ x'_A = 0$). He will then find for the coordinates of event B: ($t'_B = 1$ second, $x'_B = 225$ metres).

To summarize, we have:

	relative to aircraft	relative to ground
event A	$t_A = 0, x_A = 0$	$t'_A = 0, x'_A = 0$
event B	$t_B = 1, x_B = 3$	$t'_B = 1, x'_B = 225$
time interval between A and B	1 second	1 second
distance between A and B	3 metres	225 metres

The table shows that the two points of view agree on the time interval between A and B, but they produce very different results for the distance between A and B. In general, we find that when A and B are events at different times, it is meaningless to talk about 'the distance between A and B' unless you say how you are measuring distances—relative to what? *In classical physics, distance between non-simultaneous events is a relative concept*. This should not surprise you, as it is simply part of our everyday experience. For example, think of the events 'I fall asleep' and 'I wake up'. The distance between these two events is zero as far as the sleeper is concerned, but if you are sleeping on a train, then the distance between them relative to the ground could be hundreds of kilometres.

We see that, in general, the coordinates we assign to events can depend on how we set up the system of measurements—especially the distance measurements. This will also be true in Special Relativity, but there we have to take care with time as well as spatial coordinates.

5.2 Frames of reference

In order to discuss events and worldlines it is very useful to introduce the concept of a 'reference body', also called an *inertial frame of reference*. The idea is to try to avoid talking about 'space' and 'time' in the abstract, and instead talk about physical objects and physical change. For example, if a snail is making a journey to the end of a brick, then instead of saying 'the snail travelled 22 centimetres' we can say 'the snail travelled from one end of the brick to the other'. If meanwhile an ant scurries along the brick, then instead of saying 'the ant passed the snail at a certain point in space' we can say 'the ant passed the snail at the small bump in the middle of the brick.' Here the brick is functioning as a 'body of reference': it is a rigid object which can serve as a fixed 'background' against which various motions and events take place. By referring everything to the brick we need never mention

Figure 5.1: *Cubic Space Division*: a lithograph by Escher which captures something of the spirit of the idea of a 'reference frame'. M. C. Escher's *Cubic Space Division* © 2011, The M. C. Escher Company, Holland. All rights reserved.

the word 'space'. This is crucial, because later we are going to use physical objects to discover what 'space' is like, so we have to avoid assuming that we already know what it is at the start.

We shall also use the 'reference body' to indicate the passage of time. To do this, we have in mind that the particles of the reference body can serve as very precise and regular clocks. So instead of talking about the passage of 'time' for the snail or the ant on the brick, we could say 'between the departure events of ant and snail, the brick atom at their starting place vibrated 20 billion times'.

More generally, we would like to be able to discuss motion over large distances and in all directions, so we have to extend the idea of a 'reference body' to a huge body extending very far in all directions, but such that other things can nevertheless move through it. To get the idea, most people find it helpful to think of the reference body as a rectangular framework of thin but inflexible rods, like scaffolding, with a small clock at every intersection. This leaves room for other things to move through the framework. For this and other obvious reasons the reference body came to be called a 'frame of reference'. Do not think of 'frame of reference' as too intangible an idea. Think of it as 'framework of reference' or 'body of reference'. The idea is that the frame of reference does not have to be literally there, but we can construct it in our imagination and describe the motion of objects relative to it.

When we construct such a frame of reference in our imagination, it is important to think of it as made of real physical rods, not just abstract lines in space. However, the rods could be constructed in sophisticated ways. Any ordinary rod would be subject to undesirable limitations, such as bending under its weight, or a length depending on temperature, so we introduce an idealization: we have in mind idealized rods that represent the limit of what real rods would do if they were light and strong with no temperature dependence, and so on. We use this framework to indicate where other things are in relation to the first. For example, if the rods are all one metre long, and an event occurs half way along the third rod in the x-direction of some given frame of reference, then we say the x-coordinate of the event was 3.5 metres in that frame of reference. Also, the time at which any event occurs can be obtained by observing the reference clock situated nearest to the event.

Such a frame of reference could, in principle, be attached to any given object, so that the object does not move relative to its own frame of reference. The frame of reference then provides a way of discussing what goes on in space and time from the perspective of the object in question. Imagine the snail carrying on its back a light-weight three-dimensional scaffolding of rigid carbon fibres, with 1-centimetre spacing and extending one metre in all directions. This scaffolding, including the timing information offered by the internal vibrations of its carbon atoms, could be used to describe the motion of the ant relative to the snail. It could also be used to describe the motion, relative to the snail, of the brick or anything else.

The word 'inertial' in the technical phrase 'inertial frame of reference' is also important. It means that we are only going to use frames of reference that have constant velocity. There must be no acceleration anywhere throughout the frame of reference. If the snail on the brick changed the direction of its motion, or speeded up or slowed down, then we must imagine the frame of reference that was attached to its back continuing to move in the original direction at the original velocity, while a new frame of reference, with the new state of motion of the snail, comes into play. Of course, these two frameworks will criss-cross and bump into another, but when we appeal to a frame of reference in an argument, we mean merely that the systems of rods and clocks could in principle be present: we do not need to attach them literally to the particles or objects under discussion.

Having said that, think about the following idea. When physical entities affect one another, they never do so by 'action at a distance'. They always do it by direct local action when they are close together. For example, if a tennis racket is going to affect the trajectory of a tennis ball, it does not do so by exerting some long-range influence, but by making contact with the ball. When magnets repel one another, we sometimes feel this is a form of action at a distance, but it turns out that it is not. Rather, one magnet affects the magnetic field very close to it, that field affects the field further out, and so on, until a region of field runs up against the other magnet and exerts a force on it. Magnetic fields turn out to be real physical things, with energy and momentum. They are not just mathematical devices. Gravity can be understood in a similar way. The purpose of mentioning this fact here is to point out that the dynamics of the physical world does in fact proceed via a sort of 'physical rods and clocks' structure: all the influences propagate around by some physical means.

An example of a frame of reference is the aircraft that we considered in the golf-putt experiment. Here, the metallic frame of the aircraft provides a convenient approximation to the idealized framework we have in mind when we speak of a frame of reference. The frame of reference of the aircraft also extends throughout space, however. It is not restricted to the interior of the aircraft. It represents, at any instant of time, the whole universe, as measured by distances from the aircraft.

Another example of a frame of reference is one attached to the ground.

Our discussion of frames of reference will focus exclusively on *inertial* frames of reference. I have already said that this means frames that do not accelerate, but actually I skipped something at that point. The full definition of *inertial* is as follows.

> **Definition:** An *inertial* frame of reference is one moving either at constant velocity in the absence of gravity, or in free fall in a uniform gravitational field.

Usually we will speak of inertial frames of reference as having a constant velocity, as in the first part of this definition. However, in practice it is almost impossible to get away from gravitational fields, even in deep space, which is why the second part of the definition is needed. 'Free fall' means just that: falling under gravity, with no other forces, as a sky diver would do if there were no air. The second part of this definition turns out to be of crucial importance in General Relativity and the understanding of gravity—but it would take another book to explore it!

In the rest of this book, the phrase 'frame of reference' will always mean 'inertial frame of reference'. Also, we shall assume that the gravitational field is zero, so that inertial frames of reference have constant velocity.

We shall often want to discuss what happens, in some physical situation, using more than one frame of reference. We will mostly be interested in the case of two frames whose rods point in the same directions, but which are moving (at constant velocity) with respect to one another. One can also consider frames of reference that have a fixed rotation with respect to one another, but we will not be doing that.

By introducing reference frames we can define what we mean by speed. The phrase 'aircraft *A* has speed 500 mph relative to ground *G*' should be regarded as a shorthand for 'if you constructed a framework of rods attached to G, with spacing of 1 mile, and oriented along the direction of travel of A, then the motion of A is such that if it passes a given point in the framework when the local framework clock indicates time t, then it will arrive at a point 500 rods away when the new local framework clock indicates t plus one hour.' It seems a bit long-winded when we put it like that, but if you reflect on it you should end up agreeing that this is simply what we mean by speed, spelled out in detail. Notice that in order for the definition to work, we need to be assured that all the clocks on the framework agree with one another at any given instant: they must be synchronized.

5.3 Coordinates on spacetime diagrams

The idea of a frame of reference can be applied to spacetime diagrams as follows. We start with the simplest case: motion in only one spatial dimension. Take a diagram such as the one shown in figure 4.8. There are two particles in the spacetime region. Their worldlines are straight, so both are in uniform motion, and they each have a frame of reference. We say there is an inertial frame of reference in which the thick bead is at rest and there is an inertial frame of reference in which the thin bead is at rest.

The inertial reference frame in which a given object is at rest is called that object's 'rest frame'.

Since the tick marks on the worldlines represent the accumulation of time, we shall now indicate them by using a little clock-face symbol as shown in Figure 5.2.

Consider first the thick bead's rest frame. Think for a moment about that thick bead, or gaze at it on your wire visual aid (if you have made it), having in mind a case where the bead is at rest or is in motion at constant velocity. You should be able to see that, without a shadow of doubt, the worldline of the thick bead is a set of events *all at one spatial location in the bead's own frame of reference*. We may as well call that location $x = 0$. Therefore, the worldline

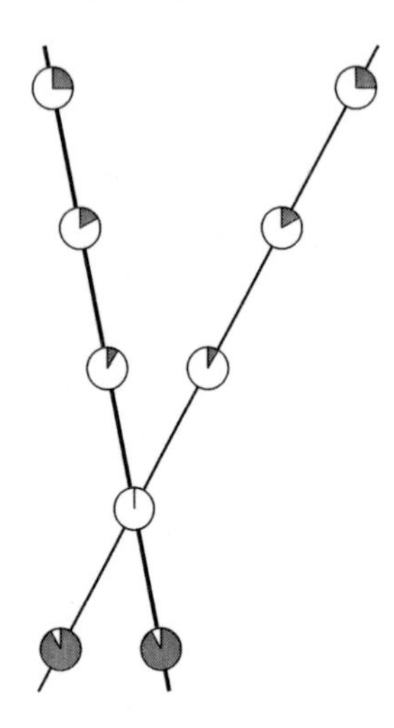

Figure 5.2: Spacetime diagram for two beads, just like the one in figure 4.8, but placing a small clock-face symbol at each tick, to indicate the passage of time.

is itself a useful coordinate indicator on the spacetime diagram: it shows the events at $x = 0$.

Now suppose there were an ideal rod attached to this bead. One end of the rod is at the thick bead, and the far end of the rod is at a fixed distance from the thick bead in the frame of reference of that bead. Imagine the far end of the rod finishes in a small sharp point. The worldline of this point must be straight, and it must be parallel to the worldline of the thick bead, because it neither approaches nor recedes from the thick bead. Therefore it appears on a spacetime diagram as a line parallel to the thick bead's worldline. By continuing this reasoning, you can see that the reference frame of the thick bead can be represented by a series of straight lines parallel to the worldline of the bead: see Figure 5.3a. Each such 'reference line' is the worldline of one point in the reference frame, and it specifies the set of events at a given spatial coordinate x in that frame.

The clock face symbols we already marked on the worldlines of the beads represent the temporal aspect of each frame of reference. In classical physics, this is quite simple, because the passage of time is the same for all inertial reference frames. We

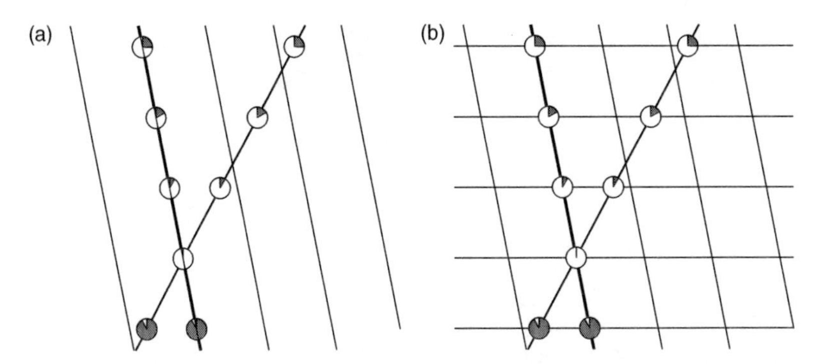

Figure 5.3: (a): Lines parallel to the worldline of the thick bead represent sets of events at fixed distance from the thick bead. (b) Horizontal lines represent successive instants of time. These 'time lines' together with the sloping 'position lines' permit the time and position coordinates, in the rest frame of the thick bead, to be identified for any given event. They function like the lines on graph paper, allowing us to determine everything from the point of view of the thick bead.

can therefore always orient the diagram so that each set of events occurring at a given time fills a horizontal line on the diagram. These lines are shown on Figure 5.3b. We can always choose one event at the thick bead to be the 'zero hour' event (for example, select midnight on 1 January 2000). This means simply that we choose to measure time intervals with respect to that event. The horizontal line passing through the 'zero time' event is the 'zero time' line, which represents all events happening at that moment. A horizontal line above this represents the set of events occurring at some later moment, a given amount of time (to be precise, a given number of clock ticks) after 'zero'. A horizontal line below it represents the events occurring at some earlier moment.

If you pick an arbitrary spot anywhere on the diagram, and would like to find at what time the event there took place, then trace a horizontal line from it back to the thick bead's worldline and see which is the nearest clock tick event.

By similar reasoning to the above, we can construct position and time reference lines for the reference frame of the thin bead. Figure 5.4 shows the construction for both reference frames

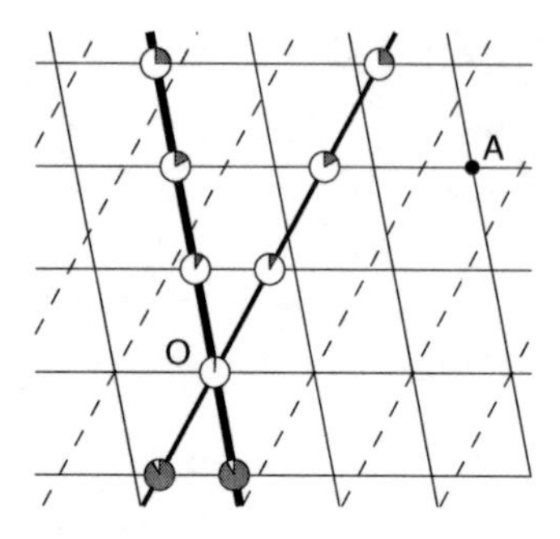

Figure 5.4: Coordinate systems for two reference frames. The bold lines are worldlines of a thick bead and a thin bead. Both have constant velocity but their velocities are different. Horizontal lines indicate sets of events simultaneous with each clock tick on the beads. Sloping lines parallel to the thick, thin bead worldline represent events at given distance from the thick, thin bead respectively. For example, event A is at a time equal to two ticks after the crossing of the beads, in both reference frames. It is at position 3 in the thick bead's rest frame, and at position 1.5 in the thin bead's reference frame. The speed of each bead in the rest frame of the other is 0.75 distance units per tick.

together. One issue requires thought, however. It is clear that the position indicator lines for the thin bead must be parallel to one another and to the worldline of the thin bead, but how far apart should we place them? Should they have the same spacing as those we have already placed for the thick bead?

In order to answer this, recall the rigid rod that was used to identify each position reference line. One end of the rod is attached to whichever bead we are studying, and at the other end is a small sharp point. Suppose we employ two rods of the same length and attach one to each bead, with the pointer end in the same direction in each case (for example, both to the right). According to classical physics and to our everyday notions of length, the size of such rods does not depend on their state of motion. For example, if they are both of length 1 centimetre, then that is the length, whether they move fast or slow or stay still. Therefore each pointer indicates the location of events that are 1 centimetre away from the relevant bead.

At the event *O* when the beads pass one another, the two rods momentarily lie one on top of the other. At the pointer end of both rods there is then an event which must be one centimetre from O in either reference frame. This tells us that the horizontal spacing between the bead worldline and the pointer worldline is the same for the two reference frames. The argument applies to pointers at each distance (for example, 1 centimetre, 2 centimetres, 3 centimetres, and so on) and thus we produce the complete picture as shown in Figure 5.4.

Notice that the position reference lines of the two frames cross one another just where they intersect the horizontal line passing through O.

The general conclusion is that in classical spacetime diagrams, the distance reference lines are separated by the same horizontal spacing on the diagram, for all reference frames, but their angles of slope differ.

As an example of the use of the reference lines, consider the event marked A in Figure 5.4. By counting the reference lines for the thick bead's rest frame, you should be able to tell from the diagram that its coordinates in that reference frame are $(t, x) = (2, 3)$. By counting the reference lines for the thin bead's rest frame, you should be able to tell from the diagram that its coordinates in that reference frame are $(t', x') = (2, 1.5)$. Also, by examining the worldline of the thin bead, you should be able to see that the speed of the thin bead in the thick bead's rest frame is $v = 0.75$.

When we come to Special Relativity, we will argue in another way by considering light signals. We will then find that the classical diagrams do not describe our universe correctly, but they are approximately valid when the speeds involved are small compared to the speed of light.

5.4 Sonar and the Doppler effect for sound

We will now apply a spacetime diagram to help us understand an interesting physical phenomenon. This will prepare the ground for Chapter 7, where we will revisit the argument and use it to unlock some central ideas in Special Relativity.

Consider two beads $\mathcal{A}$ and $\mathcal{B}$ undergoing inertial motion (i.e. uniform motion at constant velocity) along a common line, with relative velocity $\mathbf{v}$. A pulse of sound is sent out from one bead to the other, and reflected back again to the first. We would like to relate the timing of such emission and reception events to the relative velocity $\mathbf{v}$.

The use of reflected sound pulses to locate objects is widely employed in shipping (sonar) and medicine (ultrasound scans), as well as by bats. These examples are mostly concerned with measuring direction and distance, but speed information can also be obtained. Here we will just take an interest in speed measurements.

We will assume that the medium conveying the sound (such as air or water) is at rest relative to bead $\mathcal{A}$. Let the speed of sound in this medium be w. We assume the $\mathcal{B}$ bead moves relative to $\mathcal{A}$ at less than the speed of sound, so the signal from $\mathcal{A}$ can catch up with $\mathcal{B}$.

For the sake of clarity, imagine that two people, Alice and Bob, are standing on the beads: Alice on $\mathcal{A}$ and Bob on $\mathcal{B}$. Let the event when the two beads pass each other be event O. Subsequently, at event A, Alice fires a sound pulse towards Bob. Let the reflection at Bob be event B, and the final reception back at Alice be event C: see Figure 5.5. Let t_A, t_B, and t_C be the times elapsed between O and each of the other events.

First we consider the whole set of events from the point of view of Alice; that is, as observed in the rest frame of $\mathcal{A}$.

The signal reaches Bob when he is a distance

$$d = vt_B \tag{5.1}$$

away from Alice, since then Bob has been travelling for time t_B at speed v away from Alice. As far as Alice is concerned, the signal set out from her bead at time t_A, and travelled a distance d away from her at speed w. Therefore its journey took a time d/w and the arrival time at Bob must be

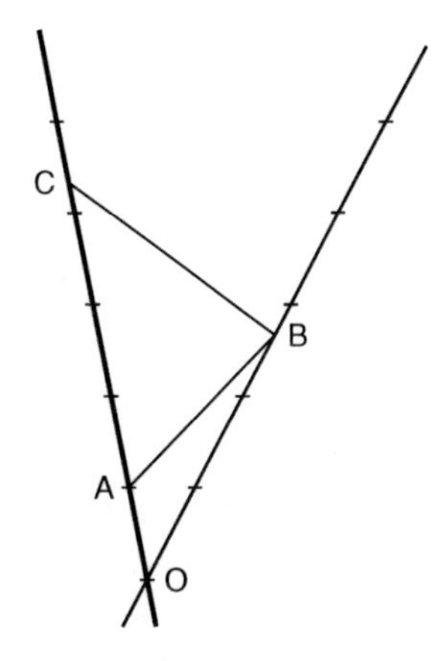

Figure 5.5: Sonar method for relative speed determination. Two beads are in constant relative motion. At event A a pulse of sound is emitted by the thick bead. It is received by the thin bead at event B, and sent back towards the thick bead, which receives it at event C. The thick bead is at rest relative to the air, and therefore the speed of the pulse relative to the thick bead is the same on the outward and return journeys. This is represented on the diagram by an equal time lapse between *A* and *B* as there is between *B* and *C*.

$$\begin{aligned} t_B &= t_A + d/w \\ &= t_A + vt_B/w \end{aligned}$$

This says that $t_A = t_B - t_B(v/w)$—$t_A = t_B(1 - v/w)$—and therefore

$$t_B = \frac{t_A}{1 - v/w} \tag{5.2}$$

On the journey back, the signal must cover the *same* distance (since Alice is not moving relative to herself), still at speed w, so

$$t_C = t_B + d/w.$$

Now use equation (5.1) again:

$$\begin{aligned} t_C &= t_B + vt_B/w \\ &= t_B(1 + v/w) \\ &= t_A\left(\frac{1 + v/w}{1 - v/w}\right) \end{aligned} \tag{5.3}$$

where in the last step we made use of equation (5.2). Make sure you are thoroughly convinced of these statements.

Equation (5.3) is the main result that we will need later. For example, it says that if $t_A = 1$ second and v is half the speed of sound, then $t_C = 3$ seconds. Note that the symbol d no longer appears in the final result: we can relate the times purely to the ratio of the two speeds.

It is instructive to do the calculation from Bob's point of view also. Bob can reason that at the time of emission t_A, Alice and her bead were at a distance $d' = vt_A$ from him. The signal had to cover this distance, and it approached him at a speed, relative to him, of $w - v$. Therefore, Bob finds

$$t_B = t_A + \frac{vt_A}{w - v}$$

Now multiply both sides of the equation by $(w - v)$, and we find

$$\begin{aligned}(w - v)t_B &= (w - v)t_A + vt_A \\ &= wt_A\end{aligned}$$

so (by dividing both sides by w)

$$(1 - v/w)t_B = t_A$$

which gives equation (5.2) as before. The reflected signal has to get to Alice when she is at distance vt_C from him. Since Bob does not move relative to himself, this is the distance it must now travel. It moves at speed $w + v$ in Bob's rest frame because Bob's bead is itself moving through the medium conveying the sound, so

$$t_C = t_B + \frac{vt_C}{w + v}$$

which, combined with the t_B equation, produces equation (5.3) once again. (Note the speed of the sound pulse *relative to Bob* is different on the incoming and outgoing journeys, but relative to Alice it is not).

Instead of firing a single pulse, Alice could emit a continuous stream of sound waves towards Bob. Suppose she does this at one particular frequency, say by singing the note 'middle C' of the musical scale, which has a frequency $f = 262$ oscillations per second. Such a sound wave has a fixed repetitive nature, repeating once every period $T = 1/f$. For example, in the case of middle C the period is 3.8 milliseconds. If we set t_A equal to this period, then we can use our calculation to find out what musical note Bob will hear. For, with this choice of t_A, the sound emitted by Alice between events O and A corresponds to one period of the sound wave, and the wavefront leaving Alice at A arrives at Bob at event B. It follows that the period between O and B corresponds to one oscillation of the sound wave received by Bob. In other words t_B is the period of the note he hears. Equation (5.2) tells us this period. For example, for the case where Bob moves away from Alice at half the speed of sound ($v/w = 0.5$), the equation tells us $t_B = 2t_A$. The period is doubled, which means the frequency is halved, so he hears a note one octave below middle C: Alice is singing like an alto, but she sounds to Bob like a bass. If Bob moves towards Alice the same equation can be used with a negative value for v, and then Bob hears a higher note than the one Alice hears. This effect is called the *Doppler effect*. It was tested in just such a musical way (but using trumpeters rather than singers) by the Dutch chemist Buys Ballot in 1845. We commonly hear the Doppler effect nowadays when a car passes by, and we hear the engine note fall in pitch. This is why children playing at racing cars like to make sounds such as 'neee-aaarr'.

Physicists usually prefer to discuss waves in terms of their frequency rather than their period. Setting $f = 1/t_A$ and $f' = 1/t_B$ for the frequencies, we find by using equation (5.2) that

$$f' = \frac{1}{t_B} = \left(1 - \frac{v}{w}\right) f \tag{5.4}$$

This is the formula for the frequency received by Bob when the source, Alice, is at rest relative to the medium carrying the sound. In the formula, w is the speed of sound relative to the medium, f is the frequency in the reference frame in which the medium is at rest, and f' is the frequency in a reference frame moving at speed v relative to the medium.

Challenge. The siren of an ambulance emits a two-tone sound whose lower note is at frequency 784 Hz. A musician knows this, and is able to identify correctly musical intervals (a change of note) to within one semitone. With what precision can such a musician estimate the speed of a passing ambulance, merely by listening to the siren? (The note one tone above 784 Hz has a frequency 880 Hz).

5.5 Conclusion

We have now finished our study of spacetime using the ideas of classical or 'Newtonian' physics. You may have begun to wonder when the 'real deal' is going to come: why all this classical stuff when we want to get to grips with the marvels of Special Relativity? The reason is that the essence of Special Relativity is a new way of understanding spacetime, so by teaching you about spacetime I have prepared the ground, while keeping your feet firmly placed on it. That is to say, by adhering to classical reasoning we have been assured that nothing odd is going to happen, and that it has all been about the familiar motions of golf balls, snails, aeroplanes, Newton's cradle, or relay runners on a track, but looked at from a spacetime point of view and a reference frame point of view. When we go on to consider Special Relativity the idea of an inertial frame of reference will be crucial. Cling on to it; it survives unchanged in the transition from classical to modern. Spacetime is also central, but in Special Relativity the way it is 'sliced up' into time and space will change.

Part 2

Special Relativity

6

The basic principles of Special Relativity

We are now ready to begin our study of Special Relativity. The 'classical' description of space and time that we studied in Chapters 4 and 5 is found to be incorrect in practice, though it remains approximately valid (to a high degree of precision) when relative velocities are small. Special Relativity is a description of motion and of spacetime that is valid whatever the speeds involved.

6.1 The Postulates of Relativity

We will study Relativity by introducing two Postulates. The word 'Postulate' is used in physics when we are going to make a 'working assumption'. This is something which we guess is right, but we are not going to try to prove it. We are just going to assume it, and then find out what would follow if it were true. The idea is that we derive as much as we can from the Postulates, and then test the whole theory by means of experimentation. If the experimental results fit the theory, then we conclude that the Postulates were valid. If we meet some sort of contradiction, on the other hand, then one or more of the Postulates must have been wrong (or our reasoning was faulty). Let me assure you that the Postulates of Relativity will turn out to be valid—even though, as we shall see, they lead to some very surprising predictions.

Here are the two Main Postulates of Special Relativity:

> **Postulate 1**, 'Principle of Relativity': *The motions of bodies included in a given space are the same among themselves, whether that space is at rest or moves uniformly forward in a straight line.*

Postulate 2, 'Light Speed postulate':
Version A: *There is a finite maximum speed for signals.*
Version B: *There is an inertial reference frame in which the speed of light in vacuum is independent of the motion of the source.*

We will discuss Postulate 1 more fully in a moment. In Postulate 2, version A, a *signal* is defined to be any means by which one physical system can influence another, in the sense of cause and effect. We will consider this point more fully in Chapter 9, where we will also prove that the two versions of the postulate are equivalent. Einstein proposed the postulate as version B. Version A is also useful because it makes it clear that the postulate does not need to refer to any specific physical phenomenon such as light or electromagnetism.

Version B of Postulate 2 refers to the remarkable behaviour of light that was described in Chapter 3. It says, for example, that if there is relative motion between the Earth and a star, the speed of light arriving at Earth from the star will be unaffected, which is exactly what is found experimentally. You might say that by introducing this as a postulate—that is, an unproven assumption—Einstein was 'cheating', in the sense that he did not offer an explanation of *why* this was so. This would be a cheat if the only thing we ever deduced from Special Relativity was that light speed in vacuum is a universal constant—but actually we can deduce much more. Einstein's genius was to see that the Light Speed Postulate was not necessarily inconsistent with the relativity postulate as had been thought, but that a radical re-thinking of the nature of *time* was needed to show how all the observations fitted together in a sensible way. The idea is that by starting with light speed as a basic idea, one can *discover* things about time (to be precise, about the rate of physical phenomena) that otherwise might seem to have no simple underlying explanation.

I chose to mention just one reference frame in version B (where it says 'There is an inertial reference frame . . .'). Often the postulate is stated in a more general way, such as 'The speed of light in vacuum is independent of the source.' I chose the more restricted statement merely in order to make the postulate as minimal as possible. The logic is that if both postulates 1 and 2 hold, then by combining them we can immediately deduce the more general result, that the speed of light in vacuum is independent of the motion of the source in all inertial reference frames. It takes some care to prove this, however, and the proof is best left until after we have become familiar with the new reasoning about space and time that Special Relativity introduces.

Classical physics is correct with regard to Postulate 1, but it does not satisfy Postulate 2. It turns out that Postulate 2 is correct and classical physics is not.

6.1.1 RELATIVITY

I have taken the statement of the Principle of Relativity from the words of Isaac Newton, because I consider his to be a clear statement. It focuses on the physical behaviour rather than on its mathematical treatment. Modern treatments of Relativity usually state the same postulate with a more mathematical flavour, such as for example 'The laws of physics can be expressed such that they take the same mathematical form in any inertial frame of reference'. This version is also helpful.

To acquire a sense of what the Relativity Postulate is all about, consider an experience such as travelling on a ship. If you have never travelled on a ship, then think of some other mode of transport such as a car, train, aeroplane, or horse-drawn cart. Now imagine conditions when the journey is smooth: the sea is as calm as a mill-pond, or the train is moving at constant speed along a smooth straight track. In such a situation you can go about your life without any need to worry about the motion of the ship (or train, car, and so on). You can read, have a drink, jump, throw a ball, all just the same as you would if you were in a room fixed to the Earth. The example of throwing a ball is a good one. If you regularly travel by car, then I encourage you to do this simple experiment (assuming you are a passenger, not the driver!) While sitting in the back seat, throw a small object such as a tennis ball

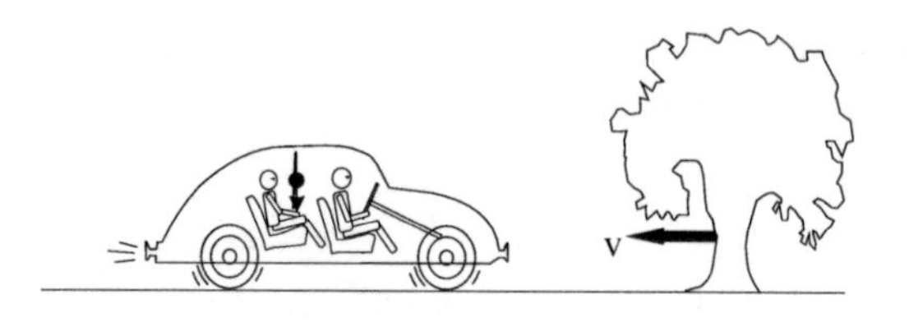

Figure 6.1: Throwing and catching a ball inside a moving car. If the car has a constant speed and direction, then it is perfectly easy to do this, because no allowance need be made for the motion of the car! For example, it is not necessary for the passenger to throw the ball forward, because *the ball is already moving forwards* (relative to the road). Inside the car, the ball simply needs to go straight up and down (and relative to the road, it traverses a long parabolic arc.) N.B. the car is moving relative to the road in *both* the upper and the lower pictures; the upper picture shows the motion of the ball relative to the car, the lower shows the motion relative to the road.

up a half a metre or so (without hitting the roof) and when it comes back down, catch it. If you can catch such a ball when sitting on a chair at home, then you will find it just as easy to catch it when sitting in a non-accelerating car. But think a little about this: do you have to allow for the motion of the car? The ball takes about half a second to go up and come back down, so if the car is travelling at 30 miles per hour, then it moved about 7 metres while the ball was in flight (see Figure 6.1). Did you have to adjust for this when throwing the ball? Did you have to throw it forwards somewhat, perhaps? Did you have to lean back to catch it?

You will find if you look into this that you do not need to make *any adjustment at all* to your throw or catch, to allow for the motion of the car. It is just as Isaac Newton said (and as Galileo appreciated before him): the motion of the ball *relative to you* is the same in a steadily moving car as it is when you are sitting in a room at home. Not even a tiny adjustment is needed. This is true as long as the car and everything in it (including the air) are all moving along together at constant velocity on a smooth road. If you want a more extreme example, take a trip on an aeroplane. But think again: you do not need to go in an aeroplane, because you are already taking a trip at 107,000 kilometres per hour 'on board' planet Earth. That is the speed of the Earth relative to the frame of reference in which the Sun is at rest. The Earth is moving at a higher speed still, relative to other stars or galaxies—and as far as our everyday lives are concerned, we could not care less.

Another way to express the idea is to say that in surroundings all moving together at constant velocity it is not possible to 'tell that you are moving' without looking at something else outside those surroundings. Some ocean liners have sports facilities such as a squash court on board. If you tried to play a game of squash in the middle of a storm at sea, you could notice that the court is tipping and shaking, but on a calm day the players in an indoor squash court on a ship will have no way of knowing whether or not the ship is moving through the ocean. The laws of physics that describe the motion of the ball, and indeed their own bodies, are completely the same in the two cases. It follows that there is no physical observation of any kind, involving objects purely inside the squash court, that could distinguish one uniform motion of the ship from another.

To take another example, think of a space ship. If there are no windows on the space ship, and there are no X-rays or other particles or gravitational fields coming in from outside, then no experiment performed inside the spaceship will be able to determine the velocity of the spaceship relative to anything outside it. If you could devise such an experiment, then you would have broken the Principle of Relativity. I am confident you will not be able to. In any case, we are going to postulate—make the assumption—that it cannot be done.

The Michelson–Morley experiment was an experiment of this kind, and it famously did not reveal any sensitivity to the motion of the Earth. According to the Postulate, tests like this should not allow even any tiny hint to be learned about a drift velocity, such as its rough size or general direction. Accelerations, by contrast, can be detected without the need to 'look outside'. For example, if a spaceship accelerates or decelerates (by firing its rocket engines), then all the matter inside will be pressed against one end or other of the spaceship. A traveller inside the ship could use such effects to deduce that the ship was accelerating relative to any local inertial frame of reference.

Is Relativity well-named?

Question: 'Is "Relativity" a good name for Einstein's theory? Classical physics obeys the Principle of Relativity, so calling the new theory by the name "Relativity" is not very helpful, as it does not distinguish it from Newtonian physics. What distinguishes the two is the Light Speed Postulate. We should use some other name.'

Answer: This is a valid point. Terminology in physics is established partly through usage as the subject develops, and it does not always succeed in being thoroughly logical. Although, along with others (including Einstein himself), I have some reservations about the name 'Relativity', it is not a bad one for the subject we are studying. Although which the Principle of Relativity is valid in classical physics, the new theory brings it to the fore and gives it much more emphasis. One of Einstein's great contributions was the *method of reasoning* that he employed in order to develop the ideas. This method involves an emphasis on the concepts of *symmetry* and *invariance*. 'Symmetry' in physics refers to the general idea that an operation performed on a system, such as rotating it or translating it, can sometimes have no discernible effect. In such a case we say the system has a 'symmetry with respect to that operation', or properties that are 'invariant under the operation'. In Relativity we meet quantities such as proper time and rest mass that are invariant under a change of reference frame, and these quantities prove to be extremely useful to help us 'find our way around' the subject. Therefore, a good name might be 'the theory of invariants' rather than 'the theory of relativity.' One name emphasizes what is absolute, the other what is relative. However, history has settled on the name we have. The word 'special' is brought in because Special Relativity is a special case of a more general theory called 'General Relativity', which deals with gravity and accelerating reference frames.

6.1.2 POSTULATES FOR DYNAMICS AND GEOMETRY

The main Postulates are sufficient to establish the basic structure of spacetime. Much of this book will be devoted to exploring this fact. When we want to study physical behaviour involving interactions between systems, however, we need a further ingredient to describe how the systems influence one another. This further ingredient can be considered a third Postulate that completes the foundations of Special Relativity. There are two ways one can introduce the third Postulate. One way is to assert that Newton's Laws of Motion hold, as long as a new formula is used to express momentum in terms of mass and velocity. We mentioned this in Chapter 2. However, we have a more profound insight if do not *assume* the Laws of Motion, but try to derive them from something more basic. We shall show that this can be done. The more basic concept is that of *conservation laws*, from which the Laws of Motion can be derived. For completeness we will state this third Postulate here, but we will postpone its further discussion until Chapter 10.

> **Postulate 3**, 'Conservation of momentum postulate': *Internal interactions among the parts of an isolated system cannot change the system's total momentum, where momentum is a vector function of rest mass and velocity.*

'Rest mass' is the mass that a body has in the reference frame in which it is not moving.

The famous equation $E = mc^2$ can be derived by careful reasoning from the three postulates: this is one of the goals of chapter 10, but there are plenty of fascinating phenomena to think about along the way.

Euclidean geometry: 'zeroth postulate'

There is another assumption which we have taken for granted in all the above reasoning, and we will continue to take it for granted throughout our discussion of Special Relativity. This is that we know what we mean by a 'distance' in any given reference frame. If one looks into this more carefully, it turns out that we are assuming certain basic rules of geometry. This can be formally noted by adding to our list the following postulate:

> **Postulate 0**, 'Euclidean geometry': *The rules of Euclidean geometry apply to all spatial measurements within any given inertial reference frame.*

'Euclidean geometry' is just a fancy name for the usual understanding of spatial measurements, where the circumference of a circle is always π times the diameter, Pythagorus' theorem can be used for right-angled triangles, and so on. I am calling this the 'zeroth' postulate because it underpins the others, and because we mostly simply take it for granted. This postulate is noteworthy chiefly because it can be abandoned in General Relativity. In the more general theory, lengths and time intervals are affected by gravity, which can lead to warping of all kinds of geometrical shapes and time measurements. We will not need to discuss this any further for Special Relativity. I have included the postulate because I promised to give you the full story, not some 'softened up' version of Special Relativity.

6.2 Simultaneity

We shall now concentrate on the two main Postulates. They are surprising because they seem, at first, to be mutually contradictory. Einstein's great contribution was to think this through with care, persistence, and rigour, and show that they are not contradictory as long as we relinquish a faulty assumption: the assumption that *simultaneity* is absolute. We are used to assuming that if two events happen at the same time, then that is all there is to be said: they are simultaneous. It turns out that this is wrong. We will find that for spatially separated events, being simultaneous is not an absolute property, because it depends on the frame of reference. Talk such as '*A* was simultaneous with *B*' is to be regarded as incomplete, because it is like saying '*A* was to the left of *B*'. Both mean nothing, or almost nothing, without further information.

Appreciating that simultaneity is a relative not an absolute concept lies at the heart of Special Relativity, and goes a long way to helping us understand the subject.

Einstein's train

Let us consider what this idea of simultaneity means. First consider an everyday example. Suppose a train is running at high speed past a station platform (see Figure 6.2). Triggers have been arranged at either end of the platform, so that when the front of the train reaches the right trigger, it causes a firecracker there to explode, leaving scorch marks on the right end of the platform and the front of the train, and emitting a bright flash of light. Similarly,

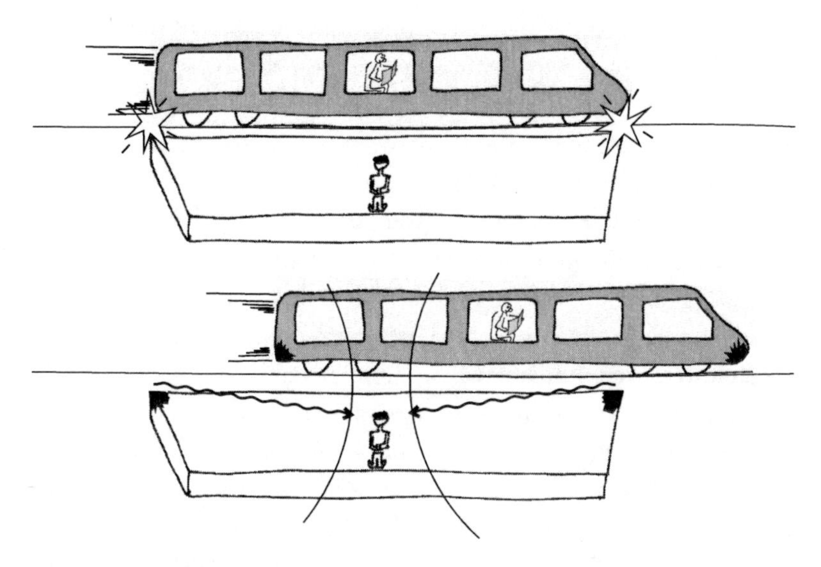

Figure 6.2: A high-speed train passes a platform. Firecrackers explode when the ends of the train are aligned with the ends of the platform. In the reference frame of the platform, train and platform have the same length, so the explosions are simultaneous. This is confirmed by the station-master, who receives both light flashes simultaneously. However, he notices that a passenger seated in the middle of the train does not perceive both light flashes simultaneously: the flash from the front reaches the passenger before the flash from the back. The passenger agrees with this, and can check from the scorch marks that the explosions were equidistant from him. He deduces that in his reference frame the explosions were not simultaneous.

when the back of the train reaches the left trigger, a firecracker scorches the left end of the platform and the back of the train, and also emits a flash of light.

The station-master is standing in the middle of the platform, and he receives the light from the two flashes at the same instant. Since he is equidistant from the two scorch marks on the platform, he deduces that the explosions were simultaneous. 'The explosions happened at the same time,' he declares.

The station-master can also examine what happens to a passenger seated half way down the train. At the trigger moment, this passenger is directly opposite the station master. While the light from the explosions is travelling, the train continues to move forward. Therefore the light flash from the front of the train reaches the passenger before the one from the back of the train. In principle the station master could confirm this by setting up fast cameras all the way down the platform, to record the progress of the bright flashes down the paint-work of the train, and examine the evidence afterwards, at his leisure.

Thus we deduce that the passenger sees one flash before the other. Perhaps the first flash arrives at him just as he is turning the page of a book, and the second flash illuminates the next page.

Now, as far as the passenger is concerned, the events of the two explosions were equidistant from him. After all, the scorch marks which they left are on the front and back of the train, and he is in a seat half way down the train. The passenger also finds that the light of the two flashes travelled at equal speeds relative to him (Speed of Light Postulate). Since the pulses arrived at him at different times after travelling equal distances, he must conclude that the explosions happened at different times. That is, in the reference frame in which the passenger is at rest, the explosions were *not* simultaneous.

'The explosions did not happen at the same time,' says the passenger. 'The explosion at the front of the train must have happened first, and then the one at the back.'

The reasoning in this example is all quite correct. The main conclusion so far is as follows.

In the example of the fire crackers and the train, two events that are simultaneous in one reference frame are not simultaneous in another reference frame.

We picked one particular pair of events for discussion, but it is not hard to see that the same conclusion follows for all sorts of other sets of events. The general idea is that the notion of 'simultaneity' cannot be given an absolute meaning: it is a well-defined notion, but a relative one (like distance and speed). It depends on reference frame. This is a remarkable idea. It means that the phrase 'at the same time' is ambiguous when applied to events at different positions. A seemingly innocuous statement such as 'the thief left by the back door just as Mr Holmes entered by the front door' has to be carefully reconsidered. Time, already famed as one of the most subtle things in our everyday experience, is even more subtle than we thought!

We do not yet have a completely general argument: it might be that there was something special about the example just presented. To understand this more fully we need to think in more general terms. We shall now do this by using spacetime diagrams.

Simultaneity on spacetime diagrams

Figure 6.3 shows a spacetime diagram with three worldlines on it. From now on all the spacetime diagrams we use will be special relativistic ones, not classical ones, so be careful not to assume too much about the diagram until we have learned more. All we are asserting so far is that the diagram has three worldlines on it: we say nothing about distance or time intervals for the moment.

Now, we are free to assert that the worldline of a particle in uniform motion (at rest in some inertial reference frame) will be straight on a spacetime diagram. This is not an assumption. It is something we can choose to guarantee by the way we define distance and time. I will not elaborate on that point here, but let us just note that it is a reasonable thing to assert, and then we can

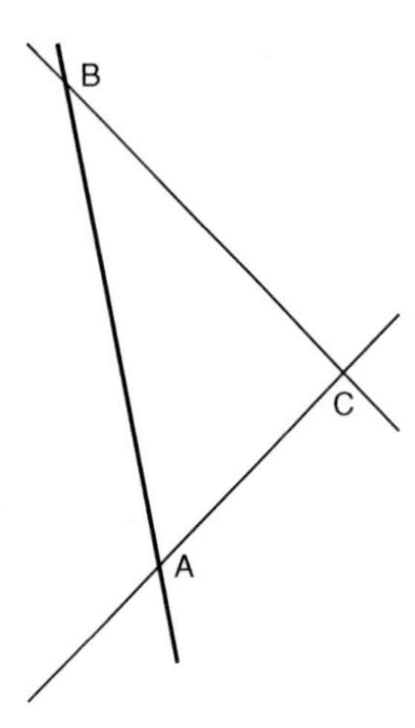

Figure 6.3: A spacetime diagram showing three worldlines, with three of the events labelled *A*, *B*, and *C*. This is a Special Relativistic spacetime diagram. We have not yet introduced the full description, and all we need for now is that the straight worldlines represent particles in uniform (constant velocity) motion. Here the two thin lines could be photons (light pulses), the thicker line something moving more slowly, such as a bead.

discover whether we can make sense of the Postulates of Relativity this way (we will find that we can).

We have now learned that the worldlines on Figure 6.3 could all represent particles in different states of uniform motion along a line. (We have a two-dimensional spacetime diagram, so we are considering motion in one dimension). Next, suppose for the sake of argument that two of the particles are travelling at the maximum speed, c. If you like, you can think of them as photons or light pulses.

Since none of the worldlines are parallel, the three velocities are different. (Same velocity implies parallel worldline: never catching up or receding. Make sure you are completely convinced of this.) If two are photons, then they must be travelling in opposite directions along the line (they have the same *speed* but different *direction*), and the third particle must be travelling at some other speed, less than c. Let us refer to this slower particle as a 'bead', to suggest that it has mass and is not a photon. Its worldline is shown thicker on the diagram.

Now 'jump aboard' the bead. The experiences of someone standing on the bead are that at event A a photon passes by to the right, and at event B a photon is received travelling back along the same line in the opposite direction (keep in mind your wire-and-beads visual aid, in order to recall that we are discussing one-dimensional motion here).

We are now going to introduce the technical term *observer*. It will be very useful to make much use of this term. The word *observer* is used to refer to the state of affairs, as a function of position and time, as found in some given frame of reference. You can think of the observer as a reasoning being standing still in the given frame of reference, but it is important not to confuse the word 'observe' with 'see'. By 'observe' we mean what the observer *deduces* must be the case, not necessarily what he or she directly saw.

First note that events *on the worldline* of an observer standing on the bead are separated purely by a time interval in his reference frame, not a distance in space, because they all occur at the same spatial location in his reference frame: namely, 'at the bead'. Clock tick events, representing the evolution of a clock at the bead, will appear on the diagram as equally spaced marks on the bead's worldline, just as in classical physics. It is clear that they should be equally spaced, because there is nothing to cause the clock to behave differently from one tick to the next, and we choose to make the diagram reflect this by using an equal spacing on the paper (or screen, or wherever your diagram is). (In technical language, we say the diagram is *linear*).

Now, the observer standing on the bead can reason as follows. A pulse of light went away from him, and a pulse came back. These pulses were not blocked by anything, so they passed each other at some position and time. Let the event where they crossed be called C. The observer would like to know what event on his own worldline was simultaneous with event C? He can easily determine this by using the Light Speed Postulate, as follows. The event C is at some distance from him in his reference frame. He does not know what distance yet, but from the Light Speed Postulate it is evident that the outgoing photon must have covered the distance in exactly the same time as the incoming photon. It follows that C occurs at a time exactly half way between A and B.

The observer is led to the conclusion that *the event on his own worldline that is simultaneous with C is half way between the photon emission event A and photon reception event B.*

If you are in any doubt about the reasoning, perhaps the following will help. Imagine the bead is sitting on a long straight path, perhaps in a nice park on a sunny afternoon. The first light pulse goes away along the path. Suppose at event C a little mirror pops up and reflects the pulse back again. Surely the light must take the same time travelling to this mirror as it takes coming back again, in the reference frame in which both the bead and the park are at rest. Therefore, the event at the bead that is simultaneous with C is the one at a time half way between the photon emission and reception.

If you have followed the reasoning so far, then you have made good progress in understanding Relativity.

Let us mark the event we just discovered, D, as on Figure 6.4. By following the same line of argument for a number of emitted and received photons, you can now prove that all the events along the dashed line marked on Figure 6.5 are simultaneous in the rest frame of the particle we are considering. To do this, just draw in further photon worldlines—which must be straight, and parallel to those we already have (because photons all travel at the same speed (Light Speed Postulate), and neither catch up with nor fall behind one another). For an outgoing photon emitted at some time before D, find the incoming photon that is received when the same time interval has elapsed after D, and where these photons cross you have an event simultaneous with D. You are strongly urged to carry out this graphical exercise. One way to do it is to use ordinary graph paper marked out in a grid of squares, but turn the graph paper through 45° so that the grid lines can act as photon worldlines: see Figure 6.5. You will find that the set of simultaneous events lies along the line shown dashed in Figure 6.5.

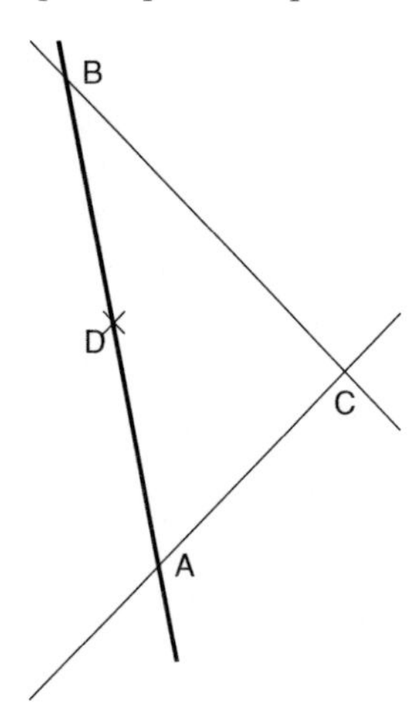

Figure 6.4: The spacetime diagram of Figure 6.3 with a further event indicated: *D* is exactly half way between *A* and *B*. An observer on the bead must find that *D* is simultaneous with *C* (in order to obey the Light Speed Postulate).

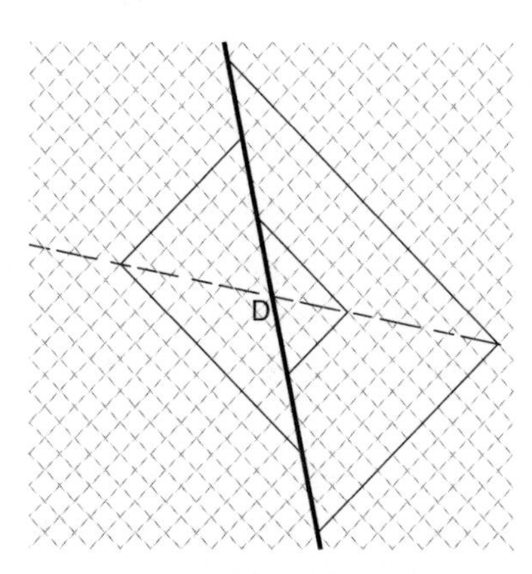

Figure 6.5: Using ordinary graph paper to construct a spacetime diagram. By turning the graph paper so that the grid lines slope at 45°, these lines can be used to represent photon worldlines. The worldline of a particle travelling at less than the speed of light is then steeper than these lines. A set of events all simultaneous in the rest frame of the particle is shown dashed. This set is found by identifying pairs of photons that are emitted and received at equal times before and after some chosen event *D* on the particle. The event where any such photon pair crosses is simultaneous with *D*. Some examples are shown.

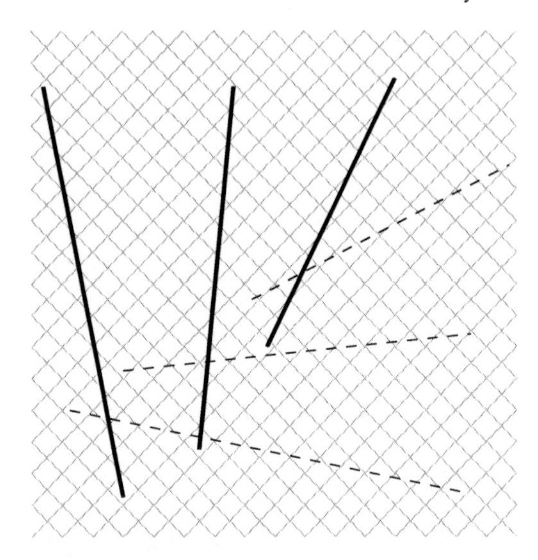

Figure 6.6: Three example bead worldlines (full), with an example line of simultaneity for each (dashed): see Figure 6.5. The mesh of 45-degree lines represents all photon worldlines. By picking pairs of these you can confirm the lines of simultaneity by using the 'radar echo' method.

After constructing the line of simultaneity for one bead, try another bead moving at some other velocity, so having a different slope on the diagram. (Make sure you always chose a speed less than the speed of light, which means a steep worldline, taking more time than a photon to cover any given distance). You will find that as you move from one bead velocity to another, the line of simultaneity changes: the worldline and the simultaneity line open and close like a pair of scissors: see Figure 6.6.

We thus discover that the lines of simultaneity for different reference frames *are not parallel!* That means, events that are simultaneous in one reference frame are not simultaneous in another. This is the **relativity of simultaneity**. Just as in the example of Einstein's train, the Postulates of Special Relativity *force* us to conclude that observers in different inertial frames of reference find different sets of events to be simultaneous.

This conclusion represents a major shift to our notion of time. Instead of time proceeding everywhere the same, irrespective of uniform motion, as is indicated by the shared horizontal lines on the classical spacetime diagrams of Chapter 4, we now have disagreement about what is simultaneous and what is not. At first sight, it may even seem that we have broken the Principle of Relativity after all. However, we have not: the notion of simultaneity turns out to be a useful construct to help us keep track of events and develop laws of motion, but when we apply those laws, the actual motions in spacetime will turn out to be the same no matter what reference frame was adopted. This is similar to the fact that the lines representing fixed *position* on classical spacetime diagrams have a slope that depends on the frame of reference (see Figure 5.4), but the predicted motions obey the Principle of Relativity.

Extending the argument to motion in two dimensions, the spacetime diagram becomes three-dimensional and the lines of simultaneity become planes of simultaneity: see Figure 6.7a. The light emitted from any given event on the worldline of a constant-

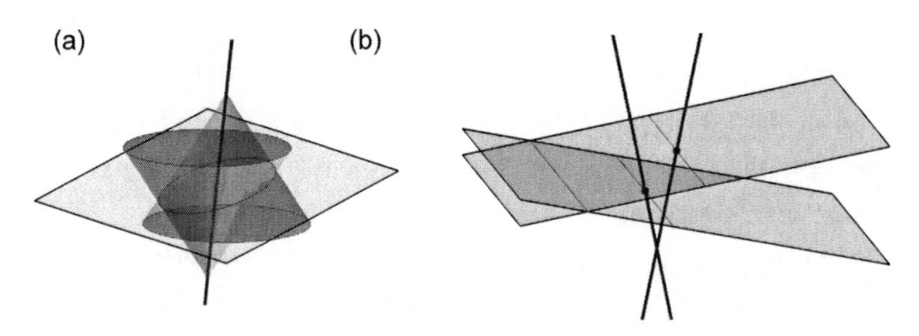

Figure 6.7: Plane of simultaneity for motion in two dimensions. (a) A particle worldline is shown, with a pulse of light propagating away from an event on the line, and a second pulse propagating towards a later event. The pulse of light looks like the surface of a cone on the spacetime diagram, because it is made of light propagating in all directions in space, all at the same speed. The events at the intersection of the cones must be simultaneous with the event half way between the emission point and reception point, in the rest frame of the particle having the worldline shown. These events lie on a plane, and further pairs of light pulses could be used to establish that the set of simultaneous events fills this plane. (b) Example planes of simultaneity for two different inertial reference frames (compare with Figure 6.6).

velocity particle now lies on the surface of a cone, and that returning to any given event lies on the surface of another cone. The cones intersect along an ellipse, lying in a single plane. That plane contains events that are simultaneous in the reference frame in which the particle is at rest. Figure 6.7b shows simultaneity planes for two different reference frames in relative motion.

For three-dimensional motion the diagram would be four-dimensional (and therefore not readily drawn by us, but we can reason about it) and would have volumes of simultaneity. We will refer to these lines, or planes, or volumes, as 'time slices': they slice up spacetime.

6.2.1 SIMULTANEITY EXAMPLE

For a numerical example of the relativity of simultaneity, suppose a pair of space-probes has been sent to the planets Neptune and Uranus. Each probe sends out a signal when it lands. If the signal from Neptune is received on Earth $1\frac{1}{2}$ hours after the signal from Uranus, then by working back, taking into account the distances travelled by the radio signals, the observers on Earth would deduce that the probes landed simultaneously, or within a few minutes of one another. Joint touchdown was at a moment approximately 2 hours and 40 minutes before the signal from Uranus arrived at Earth.

Now suppose the *same* probe landings are considered by an observer moving fast through the solar system, at four fifths of the speed of light. In the reference frame of this observer, the probe landings may or may not be simultaneous, depending on his direction of motion. For simplicity we suppose the landings occur when Uranus and Neptune are on opposite sides of the Sun, and we consider an observer moving in the direction from one planet to the other. Such an observer will find that one landing occurred about nine hours before the other. This example makes the point that the relativity of simultaneity can be a large effect. The landings were simultaneous in one reference frame, and were separated by nine hours in the other. You will be invited to check the calculation at the end of this chapter. At larger speeds or distances, the timing differences can easily extend to months and years.

If you look up at the stars, and change the direction of your gaze, then the line along which you are looking sweeps across the far reaches of space. In a similar way, if you change the velocity at which you walk or cycle, then your volume of simultaneity will sweep across spacetime (like the 'scissor blades' in Figure 6.6). However, whereas you can look in any direction in space, you cannot move at any velocity, so you cannot sweep your volume of simultaneity all the way back to the Big Bang.

6.2.2 SEQUENCE OF EVENTS: LIGHT CONES AND CAUSATION

For a given worldline, after constructing one line of simultaneity, we can easily construct others, and it is found that in the given reference frame, the sets of events that are simultaneous with successive clock ticks fill lines or planes that are parallel to one another: see Figure 6.8. This means that if we want to discover the sequence of events as time goes on in any given reference frame, we can use the method of the 'sliding slot' that we used

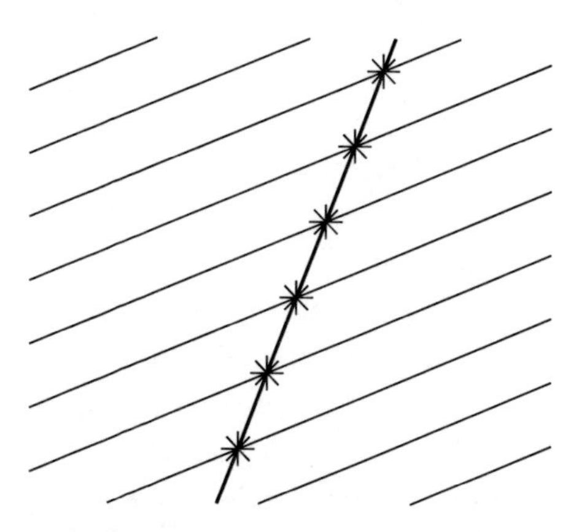

Figure 6.8: A single worldline with a set of lines of simultaneity, equally separated in time. These intersect the worldline at successive 'clock tick' events on the worldline.

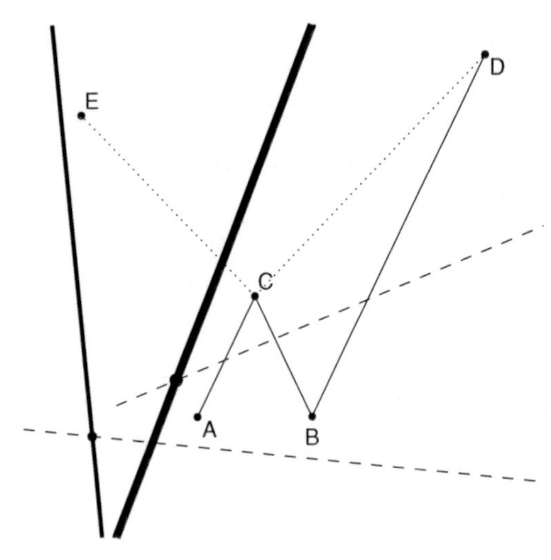

Figure 6.9: A set of events in spacetime. The two thickest worldlines represent two entities such as beads moving along a wire. Particles created at *A* and *B* propagate along the wire to *C* and *D*. At *C* two particles collide, annihilating each other and producing a pair of photons which propagate to *D* and *E*. Two lines of simultaneity are shown: one for the thin bead (dashed line of simultaneity sloping down and to the right), one for the thick bead (dashed line of simultaneity sloping up and to the right). Other lines of simultaneity for the thick bead would look similar to those on Figure 6.8; lines of simultaneity for the thin bead would also fill the diagram, all parallel to the dashed one indicated here. The simultaneity lines enable us to use the diagram to deduce the time ordering of the sequence of events. The time ordering is different for different reference frames.

for classical spacetime diagrams; but now, instead of orienting the slot horizontally we have to orient it parallel to these lines of simultaneity, or 'time slices'.

Figure 6.9 shows a set of events and worldlines in some region of spacetime. There are two bold worldlines representing two beads in uniform motion at different velocities, and various other processes. At *A* a particle appears from nowhere and propagates right, and at *B* a pair of particles is created. One of these hits the first particle at event *C*, producing two photons. One photon disappears at event *E*, and the other is absorbed by the rightmost particle at event *D*. You should now study the diagram and deduce the sequence of events, first in the reference frame of the thick bead, and then in the reference frame of the thin bead. To help you, one example line of simultaneity is shown for each of these two reference frames. Orient a slot or ruler parallel to one of these lines, then slide it up the page, keeping it parallel to the line which you selected, and thus examine the time ordering of the events according to this reference frame. Then do the same for the other reference frame. You should find that in the reference frame of the thin bead, the temporal sequence of the events is first A then B then C then E then D. In the reference frame of the thick bead, the sequence is first B, then A, then C, then D, then E.

Note that these two reference frames differ as to the time ordering of A and B, and of D and E, while they both agree that B precedes C and C precedes D. This is important, because a change in the time ordering of events introduces a danger that the whole analysis will become illogical. If B were to occur after C in some reference frame, then we would have nonsense, because event B *physically influences* event C, by sending a particle there. If B were to occur after C then we would have an effect (C) preceding one of its causes (B). For example, imagine the worldline from B to C is the worldline of a life-belt thrown to someone drowning. They catch it at C and survive. But how could they catch it before it was thrown? That would not make sense. However, our diagram does make sense because both reference frames agree that B precedes C. Similarly, it makes sense that A precedes C and C precedes D and E (C influences the latter by sending out photons). The time ordering of A and B, on the other hand, does not matter, and neither does the time ordering of D and E.

If you now look into this, you will find that Special Relativity requires different reference frames to disagree about the time-ordering of some events, but never in such a way as to lead to non-sense. In order for one event to influence another, the causal event must be close enough to and sufficiently earlier than the effect. Because signals propagate only up to a fixed maximum speed c (the speed of light in vacuum), each event can only influence part of spacetime—the part that can be reached at speeds up to *c*. However, the lines of simultaneity, or 'time slices', for different reference frames are always oriented *below*—less steeply than—the photon world lines. This is because reference frames do not travel faster than light (they can only travel as fast as real physical rods or clocks might travel). It follows that *effects always come after their causes*, no matter which reference frame you choose. If we plot the diagram so that the photon world lines are all at 45° to the horizontal on the page, then for particles whose worldlines are closer and closer to that of a photon, the line of simultaneity approaches closer and closer to a slope angle of 45°, but never exceeds it. *This guarantees that sequences of time slices never break causation*. Thus the Light Speed Postulate, which is the 'culprit' that introduced this whole business of time ordering depending on reference frame, is also the 'safety mechanism' that ensures that nonsense is avoided. This is our first example of the way in which Special Relativity introduces some notions that are surprising at first, but are not contradictory. It gives a logical and physically consistent overall picture.

Light cones

Owing to the finite maximum speed for signals (Light Speed Postulate), we can deduce that the spacetime around any given event can be divided into three regions. To illustrate this it is better to use an example with two spatial dimensions, giving a three-dimensional spacetime diagram: see Figure 6.10. Associated with any given event *A* there are two cones called 'light cones'. The sides of the light cones are photon worldlines. In two spatial dimensions, imagine a circle of light moving in towards the central event *A*, and then a flash, creating another circle of light emitted outwards.

The future light cone contains all events in the future that *A* can possibly influence. The past light cone contains all events in the past that can possibly influence *A*. The rest of spacetime contains all events that can neither influence nor be influenced by *A* (because to do so would require a faster-than-light signal). In these

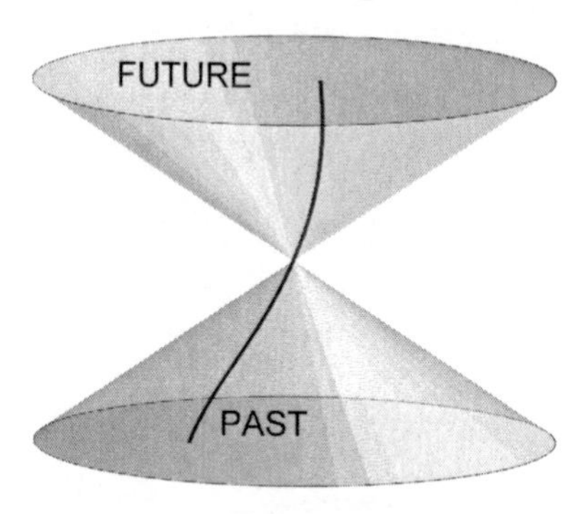

Figure 6.10: The light cones for a given event. Also shown is a worldline (not at constant velocity) passing through the event. The event can only be influenced by other events in its past light cone, and it can in turn only influence events in its future light cone. Events outside either cone are 'elsewhere': they have a spacelike separation from the event in question and can neither influence nor be influenced by it. They can be simultaneous with it.

statements *A* is an event, not a particle. Light cones provide an important basic property of spacetime. They are also greatly used in General Relativity, where gravity makes them change shape.

Light cone	Set of speed *c* worldlines passing through a given event
Past light cone	The part of the light cone in the past of the given event
Future light cone	The part of the light cone in the future of the given event
Causally connected	Events which are on or within each other's light cones
Interval	Displacement in spacetime, consisting of a time elapsed and a distance
Time-like interval	Interval between causally connected events
Space-like interval	Interval between non-causally connected events

A displacement in spacetime is a combination of a change of time and a change of position; such a displacement is called a *spacetime interval*, or just *interval*. If such a displacement lies within the light cone of one of its ends, then it is called a *timelike* interval. If it does not, then it is called a *spacelike* interval. If the interval

Figure 6.11: A worldline of a particle always lies inside the sequence of light cones associated with events on the worldline.

between two events is spacelike, we say they are not 'causally connected'.

A worldline of a particle always lies within the sequence of light cones associated with events on the worldline, because the particle always travels slower than light: see Figure 6.11. Therefore, any worldline consists of a sequence of timelike intervals.

Lines within a plane of simultaneity, on the other hand, are all spacelike. This terminology is summarized in the table.

The ring of light paradox

The relativity of simultaneity provides a neat solution to our first 'paradox' of Special Relativity. In physics the term 'paradox' is used to refer to a mistake of reasoning, from which we can learn something. A paradox occurs when two lines of argument, both apparently reasonable, lead to a contradiction. Then it is our job to find out where the reasoning went wrong, and provide a clear and correct treatment. Consider the following example. An explosion creates a bright flash of light and leaves two remnants moving at high speed in opposite directions. The flash travels outwards at

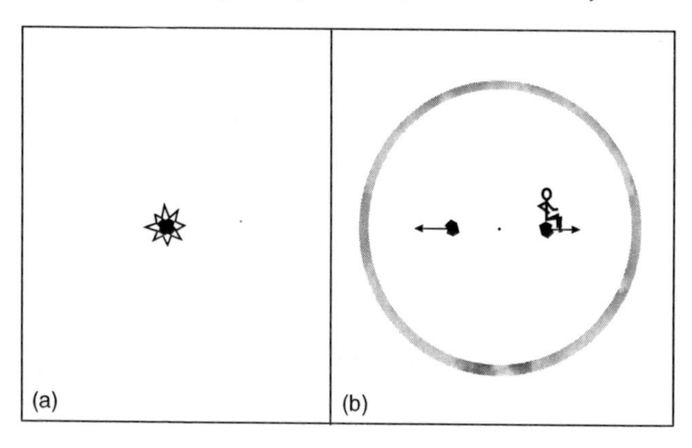

Figure 6.12: This is a picture of a situation in space, *not* a spacetime diagram. (a) An explosion creates a flash of light and some fast-moving fragments. (b) Two fragments move away in opposite directions, while the flash moves outwards at the speed of light, which means at equal speeds in all directions, so it forms a circular ring about the position where the explosion happened. The figure shows the ring, and inside it a small dot where the explosion occurred and the two fragments moving in the directions shown by the arrows. According to the Speed of Light Postulate, an observer sitting on one of the fragments ought also to observe a circular ring centred about him, because the light moved away from him at the same speed in all directions. Is the postulate wrong?

the speed of light, forming a ring centred at the position of the explosion, and the remnants move left and right. Therefore, after some time they are positioned as shown in Figure 6.12(b).

However, an observer sitting on one of the remnants must also find that a flash of light was created, and according to the Light Speed Postulate, this flash will propagate at equal speed in all directions relative to him, so he will find himself to be at the centre of the ring of light. But he plainly is not at the centre: see Figure 6.12(b). Paradox!

This paradox is resolved by the spacetime diagram shown in Figure 6.7. The reasoning given above was valid up until the last stage when we were invited to use Figure 6.12 to interpret the experience of an observer on one of the remnants. Such an observer will find that 6.12(b) *does not show a simultaneous set of events*, and this is the crucial point to resolve the paradox. The events shown on Figure 6.12(b) ('light pulse reaches here, and here, remnant is here,' and so on) do not represent the situation at any instant of time as far as an observer sitting on one of the remnants is concerned. To find the events that this observer does consider to be simultaneous, consult the spacetime diagram of Figure 6.7. In *spacetime* (not just space), the expanding ring of light traces out a cone. The straight line through the vertex of the cone can be taken to be the worldline of our observer. The plane in Figure 6.7a is then an example of a plane of simultaneity for him. The intersection of this plane of simultaneity with the light cone shows where the light flash is relative to him; that is at some instant in his frame of reference. This intersection forms an ellipse on the diagram, with the observer at the centre of the ellipse. Therefore the light has propagated to an *equal* distance in front of and behind the observer on the fragment, as measured in his own reference frame. This is precisely as required by the Light Speed Postulate. To complete the argument we need to claim that the ellipse on the spacetime diagram corresponds to a circle in space. This requires some further argument concerning the way in which these diagrams are calibrated. (To keep things simple, this further argument is postponed until Chapter 8.) Once this is done, the circle is confirmed, so the 'paradox' is fully resolved.

The story with all paradoxes in Special Relativity is that they look paradoxical when you first meet them, but once you have understood what is really going on, they begin to seem to be just a bit of faulty reasoning, and really not paradoxical at all. The aim is that they should be replaced by a smile as you see how the theory makes beautiful sense.

6.2.3 COORDINATE SYSTEMS: AN INITIAL LOOK

We can now deduce how coordinate systems are represented on special relativistic spacetime diagrams. It is not hard to see that the line representing any given spatial position in the rest frame of a given object must be parallel to the worldline of that object, just as in classical spacetime diagrams (consider a particle at such a position: its worldline must neither approach nor recede from the object). Combining such position lines with the time slices,

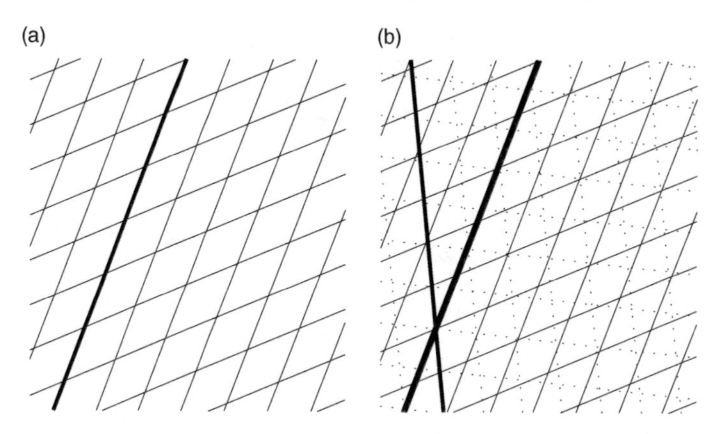

Figure 6.13: Coordinate systems in spacetime for two different reference frames, according to Special Relativity. The left figure, (a), shows a worldline of a thick bead, and the position and time coordinate system associated with it. This consists of worldlines at fixed distances from the bead, and lines of simultaneity. The right figure, (b), shows this same coordinate system, and another is shown dotted: that of a bead moving relative to the first. The diagram is introduced at this stage only in order to provide a general impression of the way spacetime is divided up into 'space' and 'time' differently for different observers. The details of how to obtain the coordinate systems are presented in the next chapter.

we deduce that the coordinate system of a reference frame will appear on a spacetime diagram as shown in Figure 6.13a, and the coordinate systems of two reference frames in relative motion appear as shown in Figure 6.13b. Compare and contrast this with Figure 5.4 on p.60. All that remains is to establish scale factors (for distance and time) of these sets of lines. The relative scales must be such that the descriptions of any given region of spacetime, in terms of one coordinate system or the other, are mutually consistent. One of the aims of the next few sections is to discover how to do that.

6.3 Time dilation

Let us now consider the passage of time. We imagine we have two beads sliding along a wire (or moving along a line in space) with

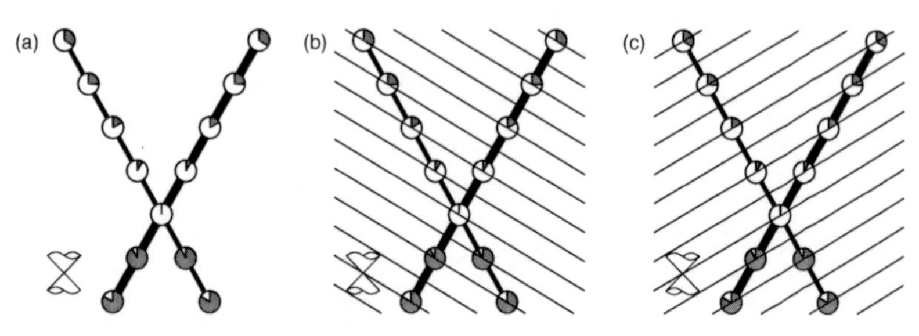

Figure 6.14: (a) Clock-tick events on two worldlines. The worldline slopes have been chosen equal and opposite on the diagram, in order to be sure that placing the tick marks at equal separations on the worldlines must represent the same type of clock for the two observers: everything is symmetric. The small cone symbol is a light cone. It acts as a reminder that this is a spacetime diagram, and it shows the orientation of the diagram and the 45-degree angle for photon worldlines. (b) Lines of simultaneity for the thin bead. It is observed that for each tick of the other bead's clock, two ticks occur at the thin bead. (c) Lines of simultaneity for the thick bead. It is observed that for each tick of the other bead's clock, two ticks occur at the thick bead. Thus in either case the moving clock takes longer to tick than does the stationary clock.

non-zero relative velocity, each equipped with a 'clock'. That is, on each bead there is a regular process of the same type going on. For example, it could be the internal oscillation of a caesium atom. (Caesium atoms have a very stable and precise internal oscillation, associated with a regular rotation of the nucleus relative to the cloud of electrons, and in fact this rotation is used to define our standard of time in 'atomic clocks'.) To represent each repetition of this oscillation—that is each 'tick'—we place a small clock-face on the worldline of each bead: see Figure 6.14a.

To be certain that the clock symbols represent the *same* type of process on the two different beads, we choose equal and opposite angles of slope for the two worldlines, and then mark the tick events with the same separation on each worldline. These choices result in complete symmetry between the two beads and the processes on them. Therefore, we can be sure we have a representation of the ticking of identical clocks carried by observers on the beads.

Now select one of the beads, and draw in a line of simultaneity at each 'clock tick' for that bead (Figure 6.14b). Look across at the worldline of the other bead, and you will see something interesting: for each two ticks of the clock on 'our' bead (the one whose rest frame we have adopted), the clock on the other bead ticks only once! Of course, the same observation happens the other way round too (Figure 6.14c). In short, each observer finds that the *other* observer's clock is 'running slow'.

The ratio came out as a factor 2 in the case of the diagram on Figure 6.14, because of the particular choice of the slope of the worldlines, which depends on the relative velocity of the beads. For larger relative velocities, the factor would be larger. For smaller relative velocities, the factor would be smaller, tending to one (equal rates) when the relative velocity is zero.

Now, this is an extremely interesting observation, for it does not matter what regular process we picked. We mentioned the vibration of an atom, because that is a very precise and regular thing, but it could have been anything: the life cycle of a tree (if our beads are as big as planet Earth), the crying of a baby, the heart-beat of an astronaut, the radioactive decay of an atomic nucleus. According to the Principle of Relativity, all these things will carry on at their normal pace in any given inertial frame of reference. But according to our spacetime diagram, when such processes are measured from a reference frame moving relative to them, they all run slow!

It is not that they merely 'look' slow—they really *are* slow. To be clear about this, it may help to picture a clock at rest in a room and another clock flying across the room at the relative speed chosen for our example spacetime diagram (Figure 6.14). Then it takes two hours of 'ordinary life' in the room for the hour-hand on the flying clock to move from the 'noon' position to the 'one' position. And the same happens when the roles are reversed.

This effect is called *time dilation*. Although the spacetime diagram contains the essence of the idea, you should not be concerned if you feel you have not fully grasped it immediately. That will come after we have looked at some more examples. Also, we have not yet extracted a formula to show precisely how the slow-down depends on the relative velocity. That will be discovered in the next section.

One example of time dilation is found in the behaviour of the fast-moving muons which arrive at the surface of planet Earth: the natural process of their break-up or decay proceeds at the ordinary pace in their own rest frame, but with their fast motion relative to planet Earth there is substantial time dilation, so that the decay proceeds much more slowly relative to the reference frame of Earth. Thus Einstein's postulates have succeeded in explaining the remarkable behaviour of muons that we examined in Chapter 3.

6.3.1 THE PHOTON CLOCK

We will obtain our next deep insight into the passage of time by using a remarkably simple argument involving a very easy-to-analyse type of 'clock'.

Our 'clock' consists simply of two mirrors fixed to a rod, with light bouncing between them: see Figure 6.15. One mirror is coated in order reflect 100% of the light incident on it. The other reflects almost all the light but allows a small amount through—say one part in a thousand—which falls on a detector. If a short pulse of light is injected between the mirrors, then as it bounces to and fro the detector will register a series of pulses. They will occur at the rate of one pulse every time interval

$$\tau = \frac{2d_0}{c} \tag{6.1}$$

where d_0 is the distance between the mirrors, because the pulse moves a distance $2d_0$ at speed c for each round trip.

You may be worried that the light intensity will fall as time goes on. Good—I hope you are, because that means you are thinking physically. But let me settle this issue. It is always possible to inject a new pulse every now and then, to maintain the brightness, and

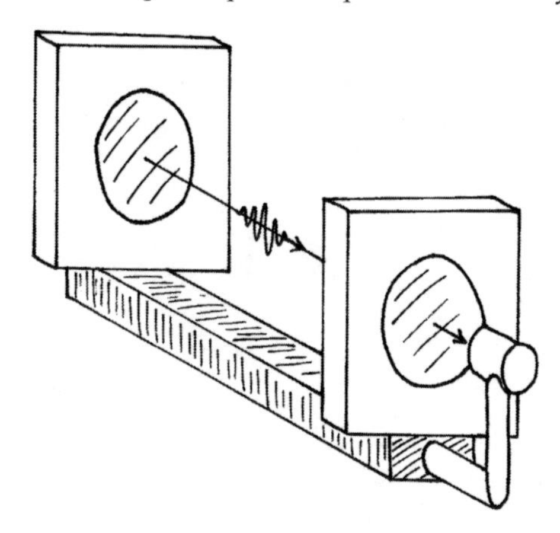

Figure 6.15: The light-pulse clock. Two mirrors are fixed to a rod. One mirror is perfectly reflecting, and the other mostly reflects but also transmits a small fraction of any light incident on it. A pulse of light is set bouncing too and fro between the mirrors. A detector (such as the human eye) near the transmitting mirror will receive a series of regularly spaced dim flashes of light. These can serve as a set of regular 'tick events' needed to mark out time. Such a clock will agree with any regular clock of any other construction, but is much easier to analyse.

by using two or more such clocks, or further types of clock, the regularity could be maintained.

Now suppose an astronaut buys two such clocks in a space station, and tests them in order to ensure that they are both in good working order, of the same size, and thus having the same 'tick rate'. Happy with his clocks, the astronaut gives one of them to his friend on the space station *S*, and takes the other with him as he boards a fast rocket *R*. The rocket moves away, accelerating up to some high speed, and then heads back towards the space station, now maintaining its velocity $\mathbf{v}$ constant. The astronaut finds his clock still works fine. In fact, if there were any problem with it, it is such a simple device that he could easily repair it himself. According to the Principle of Relativity, it must function, relative to him, exactly as it did previously in the space station. The astronaut sets up his photon clock oriented with the rod perpendicular to the line of flight of his rocket—see Figure 6.16—and enjoys

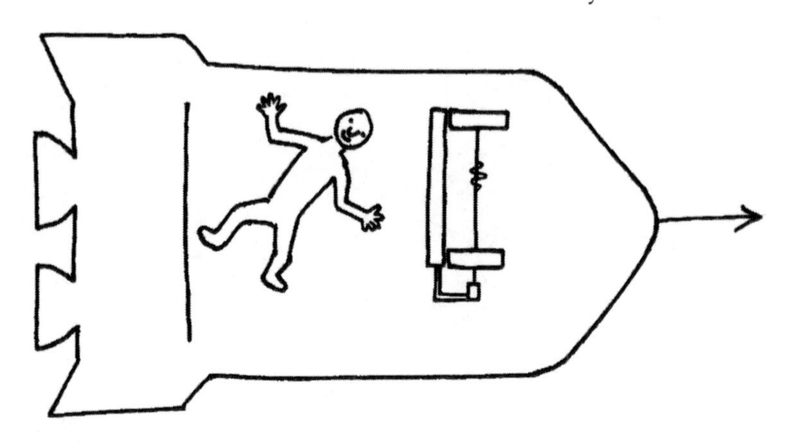

Figure 6.16: An astronaut in the rocket places his light-pulse clock next to him, oriented at right angles to the direction of travel of the rocket relative to a space-station.

listening to the clicks that the detector makes as it receives pulses of light.

As the rocket flies past her, the astronaut's friend in the space station gathers evidence about events going on inside the rocket. The sequence of events in her reference frame S is shown in Figure 6.17. As the light pulse moves inside the astronaut's clock, it travels up from one mirror to the other, but since the whole clock is in motion relative to S, the light pulse follows a diagonal path as shown. The distance the light must travel in one direction is given by the hypotenuse h of the right-angled triangle. Let $t_h = h/c$ be the time taken for the light pulse to travel this distance. Notice that by the Light Speed Postulate we are sure the light travels at speed c, so this is the correct formula. The base of this triangle has a length equal to the distance moved by the rocket in time t_h; that is, vt_h. There is also no doubt about this answer. This is, after all, what we mean by velocity: v is the defined as the distance covered per unit time.

The triangle we are considering has hypotenuse h, base vt_h, and height d_0. Now, whenever a physicist sees a right-angled triangle, she immediately has the urge to apply Pythagoras's famous theorem. So the space-station observer deduces that

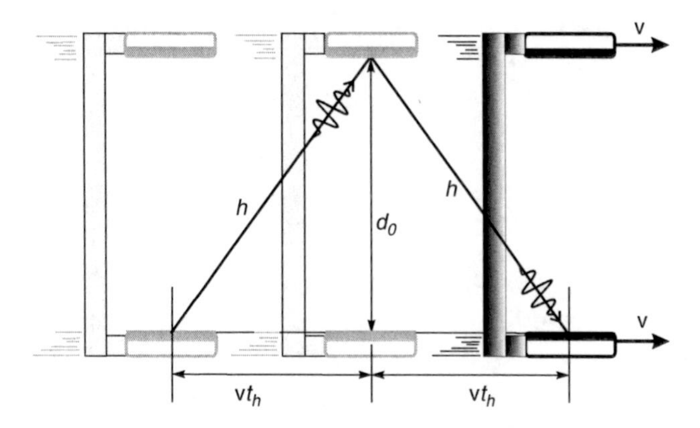

Figure 6.17: The astronaut's light-pulse clock, as observed by the space-station observer. The clock consists of a pair of mirrors with a light pulse moving between them. The mirrors are separated by distance d_0, and both move to the right at speed v. The diagram shows the position of the clock at three successive moments, as observed in the space-station. In order to move between the mirrors, the pulse of light must follow the diagonal path, as shown, covering a distance h on the way up, and another h on the way down, travelling at light-speed c. The time taken on either journey is $t_h = h/c$. Meanwhile the mirror pair moves across by a distance $v\,t_h$ for each time interval t_h. It is clear from the diagram that the light pulse takes longer to complete the round trip than it takes in the astronaut's reference frame, where it has only to travel a distance $2d_0$. By applying the Pythagorean theorem we can deduce precisely how much longer.

$$h^2 = (vt_h)^2 + d_0^2$$

But we have just argued that $h = ct_h$, so we have

$$(ct_h)^2 = (vt_h)^2 + d_0^2$$

$$\Rightarrow \quad c^2 t_h^2 = v^2 t_h^2 + d_0^2$$

Dividing by c^2 we obtain

$$t_h^2 = (v^2/c^2)t_h^2 + d_0^2/c^2 \qquad \Rightarrow \qquad t_h^2(1 - v^2/c^2) = d_0^2/c^2$$

Now divide both sides by $(1 - v^2/c^2)$, and then take the square root. We obtain

$$t_h = \frac{d_0/c}{\sqrt{1 - v^2/c^2}} \tag{6.2}$$

Therefore, the time taken for each round trip of the light pulse is

$$T = 2t_h = \frac{2d_0/c}{\sqrt{1 - v^2/c^2}}$$
$$= \frac{\tau}{\sqrt{1 - v^2/c^2}} \tag{6.3}$$

where τ is given in equation (6.1). It is the time taken for one round trip by the light pulse in the clock's rest frame, which in our case is the rocket.

For convenience, we introduce the symbol γ (as previewed in equation (2.2)), defined as:

$$\gamma = \frac{1}{\sqrt{1 - v^2/c^2}} \tag{6.4}$$

Then we have

Time dilation

$$T = \gamma\tau \tag{6.5}$$

This is the formula for the time dilation effect we first noticed in the spacetime diagram of Figure 6.14. In that diagram, the relative speed of the beads was $(\sqrt{3}/2)c$, leading to $\gamma = 2$. More generally, for a moving clock γ is always larger than 1, so $T > \tau$. This means that a moving photon clock will take longer to produce each tick than it does in the reference frame in which it is at rest—even though it remains the same clock!

This interesting result, when combined with the Principle of Relativity, has powerful consequences. The equation says that the space-station friend finds the moving clock (the one carried by the astronaut) to be 'ticking' more slowly than the clock on the space-station. But according to the Principle of Relativity, this implies that all other dynamical processes on the rocket must also be going more slowly, because otherwise the laws of motion on the rocket would be different to those holding on the space station. To prove this, we simply need to invite the astronaut to use his photon clock for the very purpose for which it was designed: namely, to time things. To be precise, let the astronaut operate the clock like a stop-watch, starting and stopping it at the beginning and end of any process he wants to investigate. For example, suppose the astronaut owns a guitar which he can play. Let him carefully pluck one string, starting the photon clock just as he releases the string (which could itself operate a switch to do this). Let him also arrange that the photon clock is stopped after one oscillation of the string. In this arrangement, the photon clock ticks some number of times. We can assert two things about this number. First, it is agreed by all observers, in whatever state of motion. If there were N tick-events, then there were N tick-events; it doesn't matter whether anyone stops to look at the place where they happened, or runs past. If you doubt that, take a more extreme example: let the photon clock break an egg every time it ticks. Then we have N broken eggs lying on the floor of the rocket: there they are, whether you fly past or stay to tidy them up. The second important property of the number N is that it is the same as the number that would be found if the experiment was carried out using a clock and guitar at rest in the space-station. If this were not so, it would mean that the internal dynamics of the guitar or the clock (or both) are different when they are at rest in one inertial frame of reference than when they are at rest in another. But that is precisely what the Principle of Relativity rules out.

Now we can put the two facts together. The astronaut carries out the experiment. Both he and the space-station observer (and anyone else) agree that his clock ticked N times while the guitar string vibrated once. But since the space-station observer finds the astronaut's photon clock to be running slow, she must also conclude that his guitar string is also vibrating slowly.

Next the astronaut tries the same type of experiment using something else: for example, a spring-based cuckoo clock. This is a clock based on coiled springs that causes a little wooden bird to come out of a box once an hour and pipe 'cuckoo'. Suppose that between each appearance of the cuckoo the photon clock ticks K times. If this K is not the same number that the astronaut previously observed when comparing them on the space station, it would mean that the internal dynamics of the cuckoo clock are different when it is at rest in one inertial frame of reference than when it is at rest in another. But that is precisely what the Principle of Relativity rules out. The space-station friend agrees that the astronaut's photon clock ticked K times, but in view of the fact that the photon clock is running slow (according to timing measurements in the space-station), she finds that the cuckoo clock must also be running slow.

You see the idea: whatever processes the astronaut has going on in the rocket, including his own heart-beat, he must find them to be in a normal time relationship with his photon clock, and this means that the space-station friend must observe them to be going in 'slow motion'. It applies to chemical reactions, to biological phenomena, to radioactive decay, to anything. But when all processes are slowed down like this, we may as well say that time itself has been slowed down: that is the summary, and the phenomenon is called 'time dilation'.

All the above arguments would also apply if we consider the astronaut's observations of the space-station clock, which is in motion relative to him: he would find that time on the space-station is slowed. These two points of view are not mutually contradictory. They merely correspond to two different perspectives on the same set of events in spacetime—take another look at Figure 6.14. Take a good long look, and reflect on the fact that it all hangs together perfectly.

As we have already noted, time dilation accounts for the remarkable behaviour of the atmospheric muons which we encountered in Chapter 3. We will treat that case fully below, but before doing so it is important to fill in some further parts of the complete picture. A full understanding, and with it a feeling of clarity and conviction, requires us to consider spatial as well as temporal phenomena, and to look at further examples.

It is interesting to notice the following implication for interstellar travel. When we consider the vast size of our Milky Way galaxy (100,000 light-years across) it may seem that one could never explore any more than a tiny fraction of it within a single human life span. As for reaching the galactic centre, some 26,000 light-years away, just forget it. . . . However, according to Relativity, an ordinary person could (in principle) go as far as you like, because a fast traveller has time slowed down: a human lifetime *on board the fast rocket* could correspond to thousands of years in the rest frame of galactic centre—long enough to get there. Such an explorer would have to accept that nothing can undo those thousands of years. They will pass for all their friends and descendants left behind on Earth, so this is somewhat like time travel into the future, without the possibility of travel back into the past.

Puzzle. It takes a human about a tenth of a second to blink. How far away from Earth is it possible to travel in the blink of an eye? (Hint: this is a trick question.)

Length of the photon clock

Before we leave the astronaut and his friend, we need to check on a point that was assumed above. This is that the length of the rod that holds the mirrors of the astronaut's clock is the same in the two reference frames. Maybe what happens is that the clock shrinks when it is in motion, so t_h in equation (6.2) is shorter, and then T could come out equal to τ: then there would be no time dilation after all. Actually, we already expected some sort of time dilation effect after noticing it on the spacetime diagram of Figure 6.14, so we know T is not equal to τ. However, it is important to look into this because if there is a change in the length of the rod then it would mean we got the wrong formula for time dilation.

We will see in the next section that rods do shrink when they are in motion, according to Special Relativity, but *not in the transverse direction*: that is, they do not shrink in the direction transverse to their velocity.

Once this is proved (in the next section) we will be able to conclude that the argument above was sound, and the time dilation equation (6.5) is correct. (And for good measure, in Chapter 7 we will derive (6.5) another way, avoiding this issue altogether.)

Puzzle. It takes one Earth-week for Robby the rocketeer, who was introduced in Figure 1.1 on p.3, to age by one day. How fast is Robby travelling relative to the Earth? (That is, can you check the number given in the caption to the figure?)

6.4 Space contraction

As already noted in Chapter 4, an object of finite size (as opposed to a particle) can be easily represented on a spacetime diagram by plotting the worldlines of all the parts of the object. For a one-dimensional object such as a stick moving along a line, we plot two worldlines for the two ends, and the region between is shaded to represent the body of the object: recall Figure 4.2. Figure 6.18a shows a spacetime diagram for two such objects in relative motion. The motion of either object is in the direction along its length, like a javelin, and we suppose they avoid collision by slipping past one another, like trains running on parallel tracks.

The relative motion between the javelins implies that their worldlines are not parallel; that is, they have different slopes on the spacetime diagram. To be precise, the worldline of the tip of one javelin is not parallel to the worldline of the tip of the other javelin. In Figure 6.18a we have made a special choice, such that the whole diagram is symmetric. This can always be done, for any pair of identical javelins, because there is always a reference frame relative to which the javelins have equal and opposite velocity. (A vertical worldline on the diagram we have drawn would represent a point in such an intermediate reference frame.) The symmetry of the diagram is useful, because it means we can claim that for a pair of javelins of identical construction, the two worldsheets must

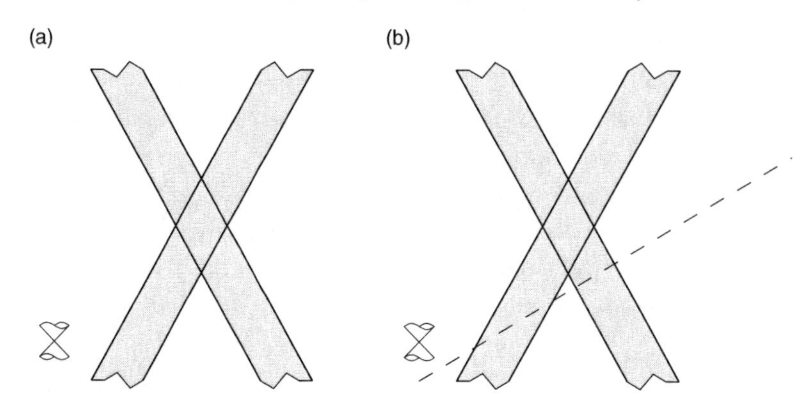

Figure 6.18: (a) Spacetime diagram showing two rods or sticks lying on the same line, and sliding along the line with some non-zero relative speed. The symmetry of the diagram implies that each rod has a length in its own rest frame equal to that of the other in the other rod's rest frame. For example, the two rods could be made the same way, of the same number of atoms, at the same temperature, and so on. Upon examining the events along a line of simultaneity for any one rod, it is seen that the *other* rod takes up less space. Figure (b) shows an example line of simultaneity for the rod moving towards the right, to help you to see the effect. Consult the main text for a complete discussion. (The rods' worldsheets are cut off at the top and bottom in an arbitrary way, and one may assume that they continue to exist at times before and after the events shown in the diagram.)

appear exactly symmetrically on the diagram, having the same width as one another, but equal and opposite slopes. The essential point is that if the javelins were to be exchanged, nothing would change. For example, if one javelin is made of 25 billion layers of aluminium atoms at room temperature, then so is the other. If one javelin is made of three hundred standard plastic bricks provided by the Lego company, then so is the other, and so on.

For brevity, let us refer to these javelins as 'sticks'. Each stick has its ordinary length in its own rest frame. An observer in the rest frame of either stick could check its length by a radar method: fire a light pulse from one end of the stick and record the time it takes for a reflection from the other end to come back. By this or other methods, you should be able to convince yourself that the two objects shown in Figure 6.18 will be found to have the same length by observers riding on them. For example, if an observer sitting on the left-moving stick finds that a light pulse takes one microsecond (millionth of a second) to travel up and back down the stick on which he is riding, then an observer sitting on the right-moving stick would also find one microsecond for such a reflection experiment on her stick. Then each observer finds his or her *own* stick to be 150 metres long. The length that an object has in its own reference frame is called its *rest length* or *proper length*.

Now choose one of the sticks, and draw in a line of simultaneity for its rest frame (recall Figure 6.6 if you need help). This has been done for you on Figure 6.18b to be sure you make no mistake. Examine the events along this line of simultaneity: you see some empty space, a shaded section representing the particles of the stick that you choose, then some more empty space, and another shaded section representing particles of the other stick. Observe that *the other stick takes up less space than the one whose rest frame you picked*. But this means it is shortened! And the effect is symmetric: either stick is shortened in the frame of reference of the other one. Furthermore, the effect arises as a direct property of spacetime. It does not depend on what sort of object is under investigation, whether a stick or a snake or anything else. *All objects are shortened compared to their rest length, when observed in a reference frame relative to which they move*. This effect is called *space contraction* or *Lorentz contraction* or *Lorentz–FitzGerald contraction* after Hendrik Lorentz (1853–1928) and George FitzGerald (1851–1901).

It does not matter what type of object is involved, and it does not matter what type of forces hold it together, whether electromagnetic and quantum mechanical (such as layers of copper atoms), or gravitational (such as a galaxy), or nuclear forces (such as an atomic nucleus or a neutron star), or whatever. The prediction of Special Relativity is that all these types of forces will be found to result in a contraction of objects when they move. When the same contraction is predicted for all types of object, we may as well say that space itself is undergoing a contraction, hence the term 'space contraction'. However, if this way of expressing it ever seems confusing, then you can always return to the idea of physical objects that become shorter.

Consider now the following example. Suppose you have a nice car: perhaps it is a classic model, or a modern super-car. In any case you would like to preserve it in good condition as you accelerate it to high speed along a road. How are you going to do that? The normal method is by rotating the wheels and using friction with the road, but this means the forces on the car are applied only at the axles. These forces will tend to distort your precious car, pushing on it more in some places than in others. To prevent this distortion, you must arrange a special machine that pushes on all the parts of the car together, at the same time. 'At the same time?' you say (as a good relativist who knows that simultaneity is not absolute), 'what do you mean? Whose "same time"? In what reference frame?' Ah! Good question. Whatever the velocity of the car, you could select the reference frame in which the car is at rest—the one travelling along with the car. But now watch what happens in the reference frame of the road: the push at the back of the car happens before the push at the front! (Take a look at Figure 6.18b again). This means the back of the car reaches the new, slightly higher, speed before the front of the car does. *Therefore the back catches up a little with the front of the car*. The result? A shorter car in the reference frame of the road!

If the forces act on all parts of the car simultaneously in some reference frame, they certainly do not act simultaneously in all reference frames. Therefore, when forces accelerate all the different parts of an object, they can't preserve its physical dimensions in all reference frames.

Puzzle. Take a look at Figure 6.19(a). This shows a stick that first moves at some constant velocity, then changes its state of motion to some other constant velocity, in such a way that it still has the *same rest length*. Choose any reference frame you like, and describe the sequence of events and the size of the stick. In your chosen reference frame, does the front of the stick change velocity at the same moment that the back of the stick does, or before or after? Does the stick become longer or shorter? Repeat the exercise for another reference frame. For example, for the reference frame moving at the initial velocity of the stick (where the stick is at rest to begin with) you should find that when the stick starts to move, it gets shorter because the back end sets off before the front. In the reference frame moving at the final velocity of the stick, you should find that when the stick comes to a halt it gets longer because the back stops before the front does. In the reference frame in which the stick maintained the same speed but simply changed direction, you should find its length is unchanged because all its parts reverse direction simultaneously.

Challenge. Now look at Figure 6.19b. Show that if all the parts of an object are given the same velocity simultaneously in the initial rest frame of the object, then the object is stretched: that is, its length in the final rest frame is longer than its length in the initial rest frame.

The spacetime diagram invites us to think of space contraction as follows. Owing to the relativity of simultaneity, the amount of space taken up by an object at any instant of time in some reference frame (what we call its length) can depend on its state of

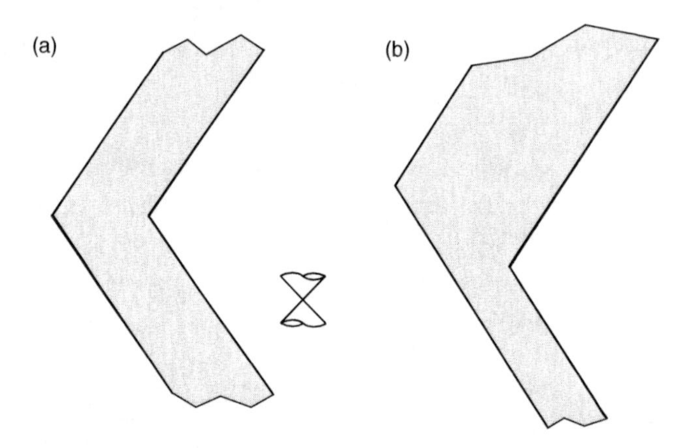

Figure 6.19: (a) A stick moves at some constant velocity, then changes to some other constant velocity without changing its rest length. Chose any reference frame and investigate the sequence of events and the length change of the stick in that reference frame. (b) The same as (a), except here the change of motion takes place in a different way: the timing is such that in the initial rest frame, the two ends change their velocity simultaneously. Now the rest length does change.

motion relative to that reference frame. This happens not because the object loses or gains any material, nor because a compressive force appears, but because the worldlines of its constituent parts (atoms or whatever) have a separation that depends on how the 'time slice' or 'line of simultaneity' picks out a set of events in spacetime to identify as 'this is the object *now*'.

6.4.1 PROOF THAT THERE IS NO TRANSVERSE EFFECT

The discussion above suggests that the contraction happens only along the direction of the motion of the object; and this is correct. We can prove this in two ways. First, consider the space-time diagram shown in Figure 6.7b. This diagram concerns motion in two spatial dimensions, so allows as to think about an object such as a flat square sliding around on an ice rink. Let two such squares have their sides aligned, and let them slide rapidly toward one another. The particles of each square have worldlines forming a rectangular tube in spacetime. A plane of simultaneity slices through this tube, producing a cross-section of rectangular shape. If your powers of visualization in three dimensions are good, then you will be able to see that the tipping of the planes of simultaneity (owing to the relative motion) only affects the dimensions *along* the line of relative motion, because the plane is angled up or down in that direction.

If you struggle to think in three dimensions, then perhaps you will prefer the following arguments.

Figure 6.20 is an illustration of one simple proof. Consider a train running very fast along a long straight track. If there were some relativistic effect causing a shrinking of an object at right angles to its motion, then in the rest frame of the railway track, the width of the moving carriage would be reduced, making each pair of wheels closer together, so that they leave the rails and hit the ground *between* the rails, gouging great ruts in the ground. But now consider the *very same situation* from the rest frame of the train. Now it is the railway track which is in motion, so the separation of the *rails* becomes smaller, so that the wheels fall

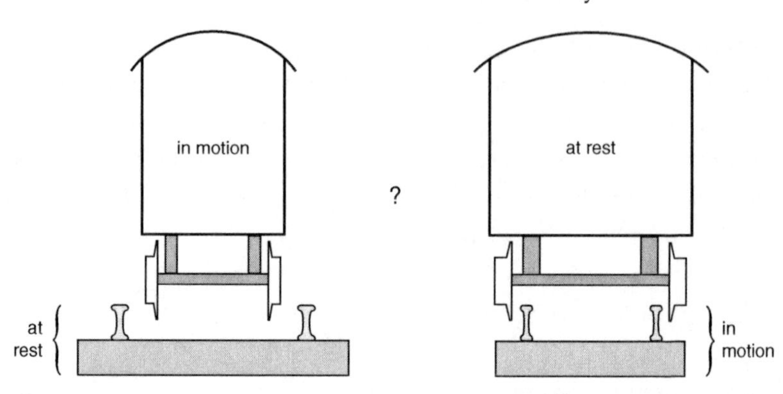

Figure 6.20: Consider the possibility of transverse space contraction. If transverse contraction happens, then according to one reference frame the fast train becomes narrower and falls inside the rails, but according to another reference frame (the trains's rest frame) the fast rails become more narrowly spaced and the wheels fall outside them. This is a contradiction, as the marks made by the wheels when they hit the ground cannot be both completely inside and completely outside the tracks. It follows that transverse contraction does not happen.

outside the rails, and dig ruts in the ground. But the marks on the ground are there for all to see: they cannot be both inside and outside. We have an impossibility, so we conclude that the shrinking does not happen. The width of the train—the dimension at right angles to its velocity—is the same when it is in motion as when it is at rest.

The same argument can be made by further examples. Here is a version that is more closely matched to our two friends with the photon clocks.

We will use the Principle of Relativity to show that when the photon clock is oriented at right angles to its motion, as it was in the discussion of time dilation in Section 6.3.1, its length is the same in both reference frames.

To prove this, let the space-station observer and the astronaut each take a rod, of identical type, which have been checked to be of the same length when they are not moving relative to one another. Now, when the rocket is moving relative to the space station, let these two rods be placed one in the space-station, and one in the rocket, both oriented transverse to the line of flight. Let each person add narrow paint-brushes to the ends of their own rod, so that the two rods paint one another as they pass. Let d_0 be the length of the space-station rod, as observed in the space-station where it is at rest. Let d be the length of the moving rod—the length of the rocket rod as observed in the space station. We suspect that d is equal to d_0, but we will not assume it at the outset, because that is what we are aiming to prove.

If the motion results in one rod becoming longer than the other, then at least one of its brushes will miss the other rod altogether as it flies past. After the rods have flown past each other, the shorter rod will be found to have one or no lines painted on it, while the longer rod will be found to have two painted lines. For example, suppose the moving rod is the same length or shorter than the stationary one; that is, $d \leq d_0$. In this case the space-station observer expects to see two marks on her rod. The two lines of paint on the space-station clock give clear evidence that $d \leq d_0$. But whatever physical law it is that results in the change in length (if one occurs), the Principle of Relativity says that the same law must apply in either reference frame. As far as the astronaut is concerned, he is at rest and the space station is moving relative to him, so he would observe the space-station clock to have the shorter rod, and it would paint twin lines on his clock. So the astronaut's clock must also have two lines on it, not one, nor none. The space-station observer, on seeing that the moving clock also has two lines painted on it, concludes that it is not shorter than the fixed one; that is, $d \geq d_0$. The only way to satisfy both conditions $d \leq d_0$ and $d \geq d_0$ is $d = d_0$.

We could not present this argument if the rods were oriented *along* their direction of relative motion. For that case we would have to arrange some sort of trigger device to fire ink-spots at the right moment, or something like that. Such an argument would have to take into account the relativity of simultaneity, and it would lead to the conclusion that we have already reached: that there *is* contraction *along* the direction of relative motion.

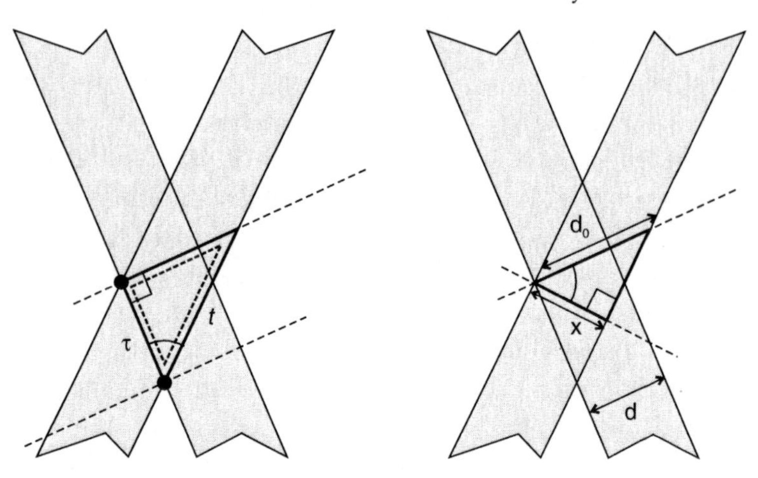

Figure 6.21: Spacetime diagram for two identical rods in relative motion, like Figure 6.18, with a geometric argument to find the amount of contraction. On the left the highlighted triangle has sides such that t and τ are the times, in the two reference frames, between the two events marked with a dot. We know these are related by $t = \gamma\tau$ (time dilation). On the right the distances marked d_0 and d are the lengths of the rods in the reference frame of the one moving to the right. Now, the triangle on the right has all its angles equal to those of the triangle of the left. This means that it must be a copy of the other triangle, but at a different size. To illustrate this, an exact replica of the right-hand triangle is shown dotted at left. It follows that its sides are in the same ratio as those of the left triangle; that is, $x/d_0 = \tau/t$. Therefore $x/d_0 = 1/\gamma$, which produces $x = d_0/\gamma$. But you can see from the symmetry of the diagram that the distance marked x is just the same as d. Therefore, $d = d_0/\gamma$. This is the formula for Lorentz contraction.

6.4.2 LORENTZ CONTRACTION EQUATION

An equation for space contraction can be obtained by a geometrical method using the spacetime diagram. Figure 6.21 shows the reasoning.

The result is that the contraction happens by the same factor γ as gives the time dilation formula:

> **Lorentz–Fitzgerald contraction** (Space contraction): the length of a moving object is shortened along the direction of motion by the factor γ, compared to its length when at rest.

$$d = \frac{d_0}{\gamma} \tag{6.6}$$

For further confirmation, consider the following simple argument. We suppose that an atom flies across a room, at speed v relative to the room. Let the length of the room be L_0 as measured in the room; that is, in its own rest frame. The time for the atom to fly the length of the room, as measured by clocks at rest in the room, is then

$$t = L_0/v$$

Now consider the fact that the atom is itself, or can be considered to be, a clock. Since it is moving relative to the room, it will not tick t times as it crosses the room, but a *smaller* number of times, owing to time dilation, given by

$$\tau = t/\gamma$$

Now consider the reference frame in which the atom is at rest. Let L be the length of the room as observed in this reference frame, and let v' be the speed at which the room moves in this reference frame. Then

$$\tau = L/v'$$

We make the following claim:

$$v' = v$$

That is, the room moves relative to the atom at the same speed as the atom moves relative to the room (only the directions are opposite). There is nothing unusual about this claim. It is just what we should expect, but I want to be clear about what is being assumed. After assuming this, we can put the above results together:

$$L = v'\tau = v\tau = \frac{vt}{\gamma} = \frac{L_0}{\gamma}$$

which is the Lorentz contraction formula (6.6) once again.

It is legitimate to ask whether the statement $v = v'$ which we made in this algebraic argument is really a separate assumption, or whether we can claim it has to be true. In fact, it follows immediately from the symmetry of the overall scenario. To deny it would amount to saying that the main postulates were not true after all. However, this is the kind of query that people like to raise when examining the main postulates in more detail; for example, by asking what would happen if light were to travel at different speeds in different directions. In the following we shall derive the Lorentz contraction formula yet another way, by considering the motion of light pulses, and doing the whole calculation in one reference frame. Since we can thus obtain the formula without having assumed $v = v'$, one could then use the above argument to confirm that $v = v'$.

The further method to obtain the formula (6.6) is somewhat more lengthy, but worth investigating. To make the argument this time we use a 'radar' method of bouncing a light signal off the far end of an object, in order to deduce the length of the object. This is none other than the photon clock once again, but now with the rod aligned along the direction of relative motion.

Consider again the scenario of the astronaut and his space-station friend, but this time let the astronaut align his photon clock *along* the direction of motion relative to the space-station (not at right angles to it as before): see Figure 6.22. First consider the astronaut's reference frame. According to his observations, the ticking rate of the photon clock, compared to other things in his rocket, shows no dependence whatsoever on its orientation. Therefore, he observes ticks separated by the time $\tau = 2d_0/c$.

Now consider the frame of reference of the space-station: see Figure 6.23. Let the length of the rod of the moving clock be d. This is the length of the rod of the rocket clock, as observed in the rest frame of the space-station. We already know the round-trip

Figure 6.22: A light-pulse clock in the rocket is now oriented along the line of its motion relative to a space-station observer.

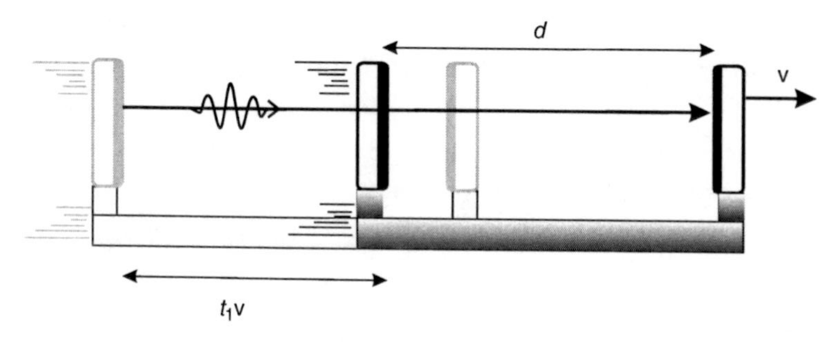

Figure 6.23: Some distances in the calculation for the photon clock aligned in the longitudinal direction. The ticking of this clock should agree with the ticking of a transverse clock, but now, according to a space-station observer, the light has to cover an extra distance in one journey between the mirrors, and a shorter distance on the way back, owing to the forward motion of the mirrors. These distance changes would not compensate for one another if the clock preserved its rest length d_0 when in motion in the longitudinal direction. A length contraction is necessary to make the timing come out right. We already know from the spacetime diagram 6.18 that such a contraction occurs (owing to the relativity of simultaneity), and this thought experiment helps us to calculate the effect precisely.

time of a light pulse in the moving clock: it is $T = \gamma\tau$. We can reuse this earlier result because the space-station observer must agree that the ticking of the astronaut's clock does not depend on its orientation. To prove this, just suppose the astronaut uses two identical clocks, one in each orientation, and sets up a tick-counter that displays the two totals. After some time has passed, the counter shows two identical numbers—say 1,000 and 1,000. Observers moving relative to the counter and the clocks must also observe these same numbers, and must agree that the counter correctly counted the ticks. Therefore, time dilation must affect both clocks by the same factor. Another way to make the argument is to consider an arrangement like the Michelson interferometer (recall Figure 3.3 on p.28). Each 'arm' of the interferometer is like a light-pulse clock. If the astronaut finds round-trip times to agree for the two arms, it means that the two halves of a light-pulse sent through the beam splitter must arrive back at the beam splitter simultaneously. Let the beam splitter itself emit a 'tick' when light hits it. Then, at the return, it emits not two ticks but one. All observers, in whatever state of motion, must agree on this, because there is only one 'tick event' for them to observe. Therefore, the round-trip time for the longitudinal clock is $\gamma\tau$, just like the round-trip time for the transverse clock.

Deriving the Lorentz contraction formula

We will calculate the round-trip time T for photons in the moving clock, in terms of times and distances in the space-station.

Let t_1 be the time taken for a light pulse to move from one mirror to the other, and t_2 be the time to come back. These times are not equal, owing to the motion of the mirrors, but $t_1 + t_2 = T$.

In time t_1, the other mirror, towards which the light pulse travels, moves a distance $t_1 v$, so the light pulse has to cover a distance $d + t_1 v$, and we deduce $t_1 = (d + t_1 v)/c$. This implies that $d = t_1 c - t_1 v = t_1(c - v)$, so

$$t_1 = \frac{d}{c - v}$$

For the returning light pulse taking time t_2, the mirror towards which the light is directed moves a distance $t_2 v$, making the distance shorter by $t_2 v$, so $t_2 = (d - t_2 v)/c$. Thus we obtain $d = t_2 c + t_2 v = t_2(c + v)$, so

$$t_2 = \frac{d}{c+v}$$

Putting it all together, the round trip time is

$$T = t_1 + t_2 = \frac{d}{c-v} + \frac{d}{c+v}$$

We are looking for a formula for d, so multiply through by $(c+v)(c-v)$ to obtain

$$\begin{aligned}(c+v)(c-v)T &= d(c+v) + d(c-v)\\ &= dc + dv + dc - dv\\ &= 2dc\end{aligned}$$

therefore;

$$\left(c^2 - v^2\right) T = 2dc$$

where we used $(c+v)(c-v) = c^2 - v^2$ to simplify a little.

To relate this to the rest length d_0, use the fact that we already know $T = \gamma\tau = \gamma 2d_0/c$. Therefore

$$\begin{aligned}(c^2 - v^2)\gamma\frac{2d_0}{c} &= 2dc\\ \Rightarrow \qquad \left(c^2 - v^2\right)\frac{\gamma d_0}{c^2} &= d\end{aligned}$$

by dividing both sides of the equation by $2c$. Now notice that $(c^2 - v^2)/c^2 = 1 - v^2/c^2$, which is none other than the inverse of γ^2, since

$$\gamma^2 = \frac{1}{1 - v^2/c^2} \qquad \Leftrightarrow \qquad \frac{1}{\gamma^2} = 1 - v^2/c^2$$

Therefore our formula reads

$$\frac{1}{\gamma^2}\gamma d_0 = d$$

which gives $d = d_0/\gamma$ as in equation (6.6).

Knowing the answer for the round-trip time, and all the speeds, we can calculate d. The method is given in the box so that you can work through it if you would like to.

Note that although this argument only produces the contraction formula for the rod in the photon clock, we can immediately assert that it applies to all objects. A spacetime diagram such as 6.18 makes this obvious. Alternatively, argue as follows. In principle, mirror pairs could be attached to, or placed alongside, all objects in the rocket, and the argument would apply to every such pair whose rod is oriented along the direction of motion. Therefore, each object must suffer the same contraction, compared to its rest length, as the one just deduced for the rod in a photon clock.

The contraction is real. It explains how the muons from the upper atmosphere can arrive at the surface of the Earth (see below), and it can be used. For example, suppose you want to paint a rocket of rest length 30 metres. The spray-painting chamber on a space-station is only 10 metres long, and for health and safety reasons it will operate only when both its doors are closed. The painting supervisor sends you the message: 'No problem: just approach the chamber at $0.97c$, and maintain a fixed velocity. Your rocket will easily fit inside the chamber. Once you are in we will do the paint job in 5 nanoseconds, and there will still be roughly 4 nanoseconds to operate the chamber doors.' One nanosecond is one billionth of a second (10^{-9} s). (Let's check the supervisor's calculation: at $v = 0.97\,c$ we have $\gamma = 4.11$, so the contracted length of the rocket is $30/4.11 = 7.29\,\text{m}$. Therefore it fits comfortably inside the $10\,\text{m}$ chamber. Once it is fully in, its nose is $10 - 7.29 = 2.71$ m from the far end of the chamber. It will cover this distance in $(2.71\text{ metre})/(0.97c) = 9.3\,\text{ns}$. This is enough time for the rapid paint job and getting the chamber doors open and closed.)

You should now feel that you understand the example given in the introduction, concerning the astronaut aboard a rocket whizzing past planet Earth. He is considered to be flat and slow by observers on Earth, while he considers Earth to be pancake-shaped and its occupants squashed and sluggish. Of course, any astronaut travelling on such a fast rocket will very likely be intimately familiar with Special Relativity, so will realize that his observations imply that Earth is roughly spherical in its rest frame, and in that reference frame its busy inhabitants are like him.

6.4.3 THE MYSTERY OF THE MUONS REVEALED

Now we can produce a thorough analysis of the amazing behaviour of the muons that arrived at the surface of the Earth. The speed of these muons, relative to the Earth and its atmosphere, is $v = 0.999c$, leading to a gamma factor of $\gamma = 22$. In their own rest frame their average lifetime is $\tau = 2.197$ microseconds (which is observed in experiments when they are not moving fast). In the rest frame of the Earth, time dilation leads to their decay taking on average $\gamma\tau = 48$ microseconds, which is long enough for a large fraction of them to survive the 50-microsecond journey to the surface of the Earth. In the muon rest frame, on the other hand, the journey time remains only 2.25 microseconds. How, then, can they reach the ground? Because in their rest frame, the Earth and its atmosphere are in motion, so that the distance the muons have to travel through the atmosphere to reach the surface of the Earth is shortened by length contraction from 15 km to about 700 metres. At a speed of $0.999c$ they soon (in 2.25 microseconds) cover this modest distance.

But does it make sense?

Question: 'You are claiming that when objects are in uniform (constant-velocity) motion, they are shorter and evolve more slowly, compared to when they are at rest. But surely they must behave precisely the same when in uniform motion, compared to when they are at rest. Is not that what the Principle of Relativity is all about?'

Answer: Nice question! But you have not understood the Principle of Relativity quite right. That Principle means that the motions of an isolated set of systems relative to a given frame of reference are the same no matter what the velocity *of that inertial frame* relative to others. However, the Principle does not rule out that observations concerning a physical object relative to a given frame of reference could depend on the velocity *of the object relative to that inertial frame*. So, for example, a car zooming along a road on Earth is shortened (compared to its rest length) in the reference frame of the road, but this contraction has nothing to do with the shared motion of both car and road relative to other things such as the planet Mars.

Notice that we only produce a complete picture in both frames of reference when we understand both time dilation and space contraction. Indeed, the same result (muon arrives in detector) is explained differently in the two frames of reference: one appeals to a temporal effect, the other to a spatial effect. This shows that in Special Relativity, space and time are no longer altogether separate or independent. Time dilation and space contraction are two sides of the same coin. In the famous words of Hermann Minkowski (1864–1909):

> Henceforth space by itself, and time by itself, are doomed to fade away into mere shadows, and only a union of the two will preserve an independent reality.

Time is different for different observers, and space is different for different observers, but spacetime is agreed by all.

CHALLENGE. Draw a spacetime diagram illustrating the propagation of fast muons in Earth's atmosphere (treat the motion in one spatial dimension). Choose for illustrative purposes a modest γ factor such as 2. Indicate events of muon creation and detection, and give a brief account of the timing and spatial considerations in the rest frames of Earth and muon.

Space contraction or physical object contraction?

Question: 'You have presented the contraction as something affecting physical objects rather than space, but it is often called "space contraction". Does it apply to the empty space between objects then?'

Answer: When we think carefully about this, it proves to be more subtle than you might guess. If you have in mind a case of two rocks hanging in empty space, with no relative motion between them, then for any reference frame with motion along the direction of a line between the rocks, the distance between the rocks will be smaller than in their rest frame. A spacetime diagram makes this obvious. I would say that to interpret this as a contraction affecting empty space tends to lead to confusion. It is better to think of it as a spatial contraction of physical objects. You might like to ask yourself how you know what the distance is between those rocks in the first place. The only way to assign some sort of meaningful distance is to come up with a physical procedure for establishing it: for example, using measuring sticks of known length, or by the timing of signals sent between the rocks. The relativistic contraction will affect whatever physical object you use, or if you use a radar method, then there will be time dilation of the physical clock. I could, for the sake of argument, take the view 'I do not need to claim anything at all about what empty space does or is like; as long as I keep track of what physical things do (including things like electric and magnetic fields) then I can provide a coherent account of everything that goes on.' The idea of empty space can be useful to our intuition, but it can also mislead. In General Relativity and quantum field theory, physicists are trying to come to grips with understanding what space and time really are, and it seems that there may be no such thing as 'empty space': everything moves against a background of fields which have discernible properties.

Let us try some example calculations of Lorentz contraction. Suppose you take a flight in a modern jet airliner. How much is the aircraft shortened compared to its rest length? Answer: Taking a typical speed as 300 metres per second (676 mph, which is approaching the speed of sound—a little fast for most airliners but a good ball-park figure) we have $v/c = 10^{-6}$; that is, the speed is one millionth of the speed of light. To calculate the contraction it helps to have an approximate formula for γ that is valid at small speeds:

$$\gamma \simeq 1 + \frac{1}{2}\left(\frac{v}{c}\right)^2 \tag{6.7}$$

(We shall derive this in Chapter 10.) In this example we find the contraction is hardly noticeable: γ differs from 1 by only 0.5×10^{-12}, and therefore the length of a 60-m aircraft changes by about 0.03 nanometres, which is less than the diameter of one atom. This is so small that the merest brush with the air in front of the aircraft would have a much larger effect.

In search of a more noticeable contraction, you have to consult particle physics or astronomy. Consider, for example, the planet Mercury. Its mean diameter is about 4,880 km in its rest frame, and its average orbital speed around the Sun is 48 km/s. The Lorentz contraction in this case reduces the diameter in the direction of motion by 63 centimetres—not exactly a big deal, but large enough to be noticeable in a careful survey.

Puzzle. For a more impressive result, consider the atmospheric muons again, whose motion relative to the Earth results in a Lorentz factor $\gamma = 22$. What is the shape of planet Earth in the rest frame of such a muon?

6.5 Conclusion so far

In this chapter we have introduced the Main Postulates of Special Relativity, and we have shown how they lead to definite predictions about the lengths of physical objects and the rates of repetitive processes such as those going on in physical clocks. Hence they lead to predictions about the rates of all types of processes. An excellent insight into these predictions, and especially the way they mesh together without contradiction, is provided by the concept of spacetime and the use of spacetime diagrams. For example, Figure 6.14 makes it clear that each of a pair of observers can find that the *other* one is 'ticking more slowly', without contradiction.

Historically, Special Relativity was discovered at a time when the evidence for a break from classical (Newtonian) physics was mounting. This is the evidence outlined in Chapter 3—especially the evidence from optical experiments and the theoretical study of Maxwell's equations. Einstein was not the only one to make notable contributions, but he did carry the crown because he made the crucial breakthrough. A summary of the main contributors is shown in the box.

In the next chapters we will discover further features of the 'relativistic world'—which is none other than the world we inhabit. This will involve looking at time dilation and space contraction again from other perspectives, and examining more examples of what they mean in practice. After that we will turn to the study of momentum and energy, and the crowning achievement of Relativity theory—the equivalence of mass and energy—will emerge.

Puzzle. Using a straight edge, measure the spacetime diagram shown in figure 6.24, in order to confirm the statements presented in Section 6.2.1 concerning the timing of two probe landings on Uranus and Neptune.

Challenge. *If you can complete this challenge, it will make you feel considerably more comfortable about all the ideas introduced in this chapter.* The relative speed of the train and platform illustrated in Figure 6.2 is $v = (3/5)c$. The train and platform have the same length of 20 metres in the rest frame of the platform. The speed of light is 3×10^8 metres per second.

(i) First consider the rest frame of the platform and station-master. Show that the light pulses take 33 ns to reach the station-master. Show also that the front pulse reaches the centre of the moving train 21 ns after the explosion, and the back pulse reaches the centre of the moving train 83 ns after the explosion. Note that the gap between these two times is 62 ns.

(ii) Now think about lengths. Determine the rest length of the train, and the length of the platform in the rest frame of the train. Hence show that in the rest frame of the train, its length exceeds that of the (moving) platform by 9 metres.

(iii) Using the answer to part (ii), explain the sequence of explosions in the rest frame of the train. Which explosion happens first? What time interval passes between the two explosions?

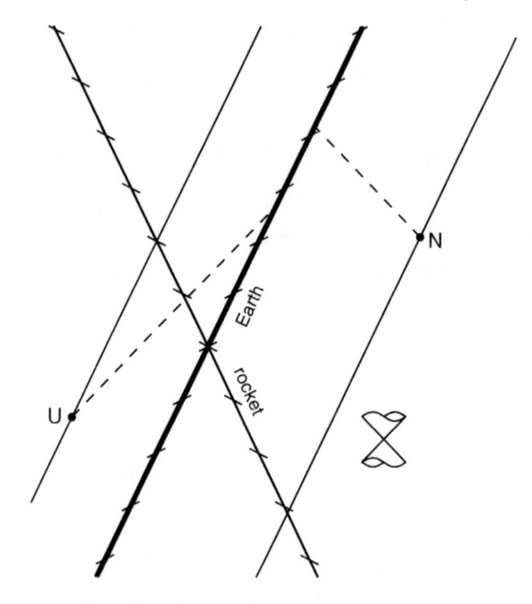

Figure 6.24: Spacetime diagram for the simultaneity example presented in Section 6.2.1. Full lines are worldlines of planets and the rocket; dashed lines are radio signals. The relative motion of the planets is negligible on the scale of the diagram. The probe landing events are marked U and N. In the reference frame of Earth they are simultaneous; in the reference frame of the fast rocket they are not. You should now know enough about these diagrams to be able to measure this one with a ruler and confirm the statements presented in Section 6.2.1. The tick marks on the worldlines of Earth and the rocket are at intervals of 1 hour.

Pioneers of Special Relativity

James Clerk Maxwell completed the foundations of electromagnetism, and showed that light and radio waves are electromagnetic waves propagating at a fixed speed given by his equations.

Hendrik Lorentz performed many penetrating analyses of the implications of Maxwell's equations, and in particular realized that they implied that material bodies held together by electromagnetic forces would shrink along their direction of motion. He discovered the important 'Lorentz transformation' equations in the context of electromagnetism.

Albert Michelson and **Edward Morley** performed crucial experiments to search for an influence of the motion of the Earth on the speed of light.

J. J. Thomson and **Max Abraham** suggested from electromagnetic theory that charged bodies would show a contribution to their mass that varied with velocity, and **Walter Kaufmann** gained experimental evidence for such a variation.

Henri Poincaré discussed many of the principles that underlie Special Relativity, and came close to discovering it. His work completed and extended that of Lorentz, and it included all the mathematical tools subsequently adopted by physicists to study Special Relativity. However, he left unclear the simplicity of the fundamental postulates and the fact that the aether concept was not needed.

Albert Einstein in 1905 made the foundations of the subject clear and concise. In particular he showed that the behaviour of moving bodies could be developed from simple postulates through a radical reinterpretation of the meaning of time.

Hermann Minkowski developed the great insights obtained by treating the whole subject in terms of four-dimensional 'spacetime'.

Others worthy of mention include **Fresnel**, **Heaviside**, **Hertz**, **Fitzgerald**, and **Voigt**. This list of names and the associated ideas show that Relativity did not emerge in a vacuum. Rather, as Einstein himself acknowledged, Special Relativity was 'ripe for discovery in 1905.' Nonetheless, a new and penetrating way of thinking was needed, and for this central contribution Einstein is rightly credited.

(iv) Hence find the time experienced by the passenger between the two events at which the two light flashes reach him.

(v) Explain how time dilation relates the answer of part (iv) to the 62 ns found in part (i).

7

Foundations re-explored

In this chapter we will discuss a simple set of events all in one spatial dimension. Among other things, we will derive the time dilation formula (6.5) again: it is reassuring to be able to deduce it in more than one way. The argument is very 'clean', in that we do not need to worry about transverse effects, and we try to rely on the minimum of basic statements about time and motion. By careful application of the two Main Postulates of Relativity, we will produce, almost as if by magic, several 'golden nuggets' of knowledge: the time dilation factor, the Doppler effect (change of frequency due to motion) for light, and the relativistic formula for adding velocities.

We want to revisit the idea of time proceeding differently in different reference frames, but once again without introducing *time* as such: we concentrate instead on some regularly repeating physical process. That is, we talk about 'clocks' rather than 'time'. Now, a given clock only directly indicates the passage of time *right there at the clock*. We need to be careful if we also use it to measure events happening somewhere else, at some distance from the clock. This issue was present in the background of the discussion of the photon clock in Section 6.3.1. There we talked about 'the time it takes for the light to perform a round trip' without worrying about the fact that, for a moving pair of mirrors, the round trip ends at a different place from where it started (it ends at the mirror, but the mirror has moved). This was acceptable, because we had already carefully defined simultaneity for events at different places. In this chapter we present a clever argument, due to Hermann Bondi, that has fewer hidden assumptions. It has a disarming simplicity, but in some ways it is more subtle than the arguments presented in the previous chapter, which is why I have presented it second. In any case, it is always useful to have more than one line of argument for anything basic in science: each line has its own merits.

We consider a scenario just like the one discussed in Section 5.4 on sonar and the Doppler effect for sound, but now instead of sound we use light waves, and we carefully employ the postulates of Special Relativity.

Suppose there are two objects that evolve in some regular way, and which are identical: for example, two caesium atoms. Let one such atom be held by Alice on bead S and one by Bob on bead S′. The beads move along a line and exchange signals just as described in Section 5.4, only here the signal is a pulse of light rather than sound. The spacetime diagram which was previously as in Figure 5.5 is now as shown in Figure 7.1.

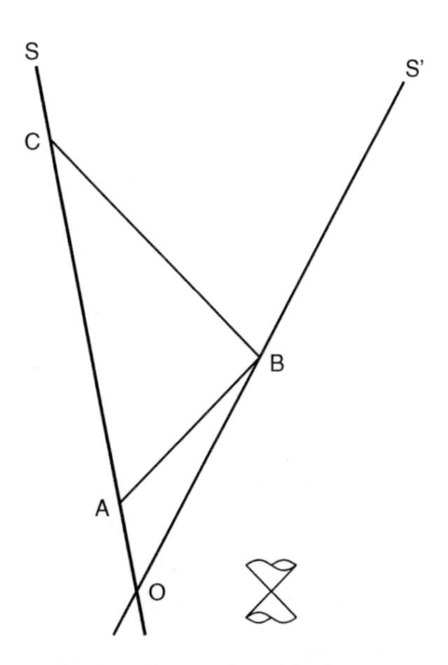

Figure 7.1: Spacetime diagram for a radar echo between two beads in relative motion along a line. The time dilation formula, and other results, will be obtained by argument directly from this scenario. The diagram has been constructed by giving the photon worldlines equal and opposite slopes.

Let t_A, t_B, t_C refer to the *number of oscillations of Alice's clock* between O and events A, B, C respectively. For t_A and t_C the oscillation counts or 'ticking numbers' concern events directly at Alice's clock, so no further comment is needed. For t_B, however, we have to say what we mean, because the clock is held by Alice but the event B is at Bob. We define t_B to be the oscillation count between O and the event on Alice's worldline that is considered by her to be simultaneous with B. We have already established (in Section 6.2 on simultaneity) that since it is a light signal that is travelling to and fro, we must find that such an event is half-way between A and C. That is,

$$t_B = \frac{t_C + t_A}{2} \tag{7.1}$$

Let t'_B refer to the number of oscillations of Bob's clock between O and B. This is a count of events directly at Bob's clock so needs no further comment. (We could go on to define t'_A and t'_C but we will not need them.)

The reasoning that was given for a sound pulse in section 5.4 carries over directly to a light pulse, as long we use the reference frame where the sound pulse had the same speed on both journeys. This means that equation (5.3) can be applied directly to the case of light waves and our clock oscillation counts:

$$t_C = t_A \left(\frac{1 + v/c}{1 - v/c} \right) \tag{7.2}$$

You are strongly encouraged to revisit that argument now, to be sure that it applies to a signal propagating at the speed of light, as claimed.

Next we take a clever step, by noting a certain symmetry in the physical scenario. Events B and C have the same character as events A and B. In both pairs of events, one observer sends out a signal, and the other observer moving away at speed v receives it. By the Light Speed Postulate, the signal propagates away from the sender at the same speed in both cases. By the Relativity Principle, the same equation must relate sending and reception times in the two cases. Note that this symmetry is lacking in the case of sound waves, because there the speed away from the sender depends on the motion of the sender relative to the medium of propagation of the sound. Light needs no such medium, and has the same speed in either case.

By the Main Postulates, therefore, we can define a 'ticking ratio' $R = t'_B/t_A$ as 'the number of receiver clock ticks divided by the number of sender clock ticks, where the counts start when two clocks pass each other, and finish at the events when a given light signal is received or sent, respectively, the clocks being of identical type and moving uniformly at relative speed v.' This definition can apply equally to t'_B/t_A or t_C/t'_B, so we must have

$$R = \frac{t'_B}{t_A} = \frac{t_C}{t'_B} \tag{7.3}$$

It is important to think carefully about this equation. Much of Relativity is contained in it!

Now we already know the ratio of t_C to t_A (equation (7.2)), and we can express

$$\frac{t_C}{t_A} = \frac{t_C}{t'_B} \times \frac{t'_B}{t_A} = R \times R \tag{7.4}$$

by using (7.3). Therefore, after using (7.2) to express t_C/t_A, we have

$$R = \sqrt{\frac{1 + v/c}{1 - v/c}} \tag{7.5}$$

This is the end of our derivation. We have a formula for the ratio R of reception time to sending time, when light pulses are sent between parties.

7.1 Golden nugget 1: time dilation

Now divide equation (7.1) by t'_B, and use $t'_B = Rt_A$ (from equation (7.3)) on the right-hand side:

$$\frac{t_B}{t'_B} = \frac{t_C + t_A}{2t'_B}$$

$$= \frac{t_C + t_A}{2Rt_A} = \frac{R^2 t_A + t_A}{2Rt_A}$$

where we used (7.4) to replace t_C. Therefore,

$$\frac{t_B}{t'_B} = \frac{t_A(R^2 + 1)}{t_A(2R)}$$

$$= \frac{R^2 + 1}{2R} \tag{7.6}$$

$$= \frac{1}{\sqrt{1 - v^2/c^2}} \tag{7.7}$$

where the last step requires some thought (see below). The result is $t_B = \gamma t'_B$. This is the time dilation effect (compare with equation (6.5)), already discussed in Section 6.3, but now we have derived it by an argument involving motion in only one spatial dimension, so we did not need to consider the transverse dimension of any object. Also we made very clear where the concept of simultaneity enters (eq. (7.1)) and where the Main Postulates were invoked (equation (7.3)). In several ways this is a cleaner argument than the one leading up to equation (6.3) on p.102. In any case, both arguments are useful: especially their agreement of course!

Mathematical Challenge. Derive equation (7.7) from equation (7.6). (Hint: first show from (7.5) that $R^2 + 1 = 2/(1 - v/c)$; after that, you can if you wish square the whole equation to help simplify it, and then take the square root at the end.)

7.1.1 PROPER TIME

The best way to think of the time dilation result is to consider it as an equation relating the time between two events to the 'proper time' between them:

Definition. The *proper time* between two events is the time interval between them as recorded in a frame of reference in which they both occur at the same place.

We can think of objects as travelling through space carrying their 'proper time' with them, since all the events in a given object obviously occur at that object. Each object (atom, bacterium, whatever) behaves in an ordinary way as measured by its proper time. When measured by an observer moving with respect to the object, the time intervals between events on its worldline are longer.

7.2 Golden nugget 2: relativistic Doppler effect

The 'ticking ratio' R describes the Doppler effect for light. The argument is essentially the same as the one we gave in Section 5.4 for sound waves in classical physics. For, if Alice emits pulses in a regular sequence, once every time t_A, then it is easy to see that they will be received in a regular sequence, once every time t'_B, because there is nothing to distinguish one pulse in the sequence from another, and the relative motion is at constant velocity. Now suppose each pulse is in fact a peak of a continuously emitted set of waves of fixed frequency, then t_A and t'_B are the periods of the waves sent and received, so the frequencies are related by

$$\frac{f'}{f} = \frac{1}{R} = \sqrt{\frac{1 - v/c}{1 + v/c}} \tag{7.8}$$

where f is the frequency transmitted, f' is the frequency received, and the receiver moves away from the transmitter at relative speed v.

Equation (7.8) describes the Doppler effect for light, sometimes called the 'relativistic Doppler effect'. The received frequency is *lower* when the receiver is moving *away* from the source. This is as one would expect. Lower frequency corresponds to longer wavelength, which in the visible spectrum means a shift towards the red wavelengths, so we speak of a 'red shift'. When the receiver moves towards the source, there is a shift upwards in frequency, called a 'blue shift'. The main point, however, is that we now have a precise formula for the expected shift, not just a rough intuition about it. This formula has been tested in vast numbers of experiments in atomic physics, where the frequencies of light-waves emitted by atoms are very well known, and thus repeatedly confirmed by experiment. This in turn permits some very important insights in astronomy.

To the naked eye, at first sight the stars all appear to be white in colour. However, if you look more carefully you will see that this is not quite true: some have a more bluish tinge, and some red. This is more easily seen using binoculars or a telescope. By a careful analysis of the colour of the light received from any given star, astronomers can deduce much about the star. For example, the colour differences noticeable using binoculars are largely owing to different surface temperatures of the stars. More precise observations reveal that for almost all stars there are very narrow bands of colour that are missing. This can be understood: the atoms in the outer layer of the star are absorbing some of the light. The pattern of these missing bands is a sort of 'signature' of the atoms, and in fact the chemical element helium was first discovered this way in the outer layers of the Sun, before it was found on Earth! In the distant stars these atomic signatures are readily recognised, but when the stars are in galaxies other than our own, the missing bands are found at the 'wrong' places, not corresponding to any known atom. That is, they do not correspond until one allows for a shift in frequency of all the colours. With such a shift, the patterns all match perfectly the ones we find in studies of atoms on Earth. This shift is interpreted as a Doppler effect, and it shows that the galaxy in question is moving away from our own. The more distant galaxies are found to be moving away more quickly. For the most distant galaxies imaged by the Hubble Space Telescope, one does not need a sophisticated analysis: the light looks distinctly red even to our rough human colour sense. This sort of motion of all the galaxies represents a uniform 'swelling' of the whole of space. It is called the expansion of the Universe. It is one of the great scientific discoveries of the twentieth century.

CHALLENGE. The function on the right-hand side of (7.5) arises quite often in professional work with Relativity, and you may like to confirm that it can also be written

$$\sqrt{\frac{1+v/c}{1-v/c}} = \frac{1+v/c}{\sqrt{1-v^2/c^2}} = \gamma(1+v/c)$$

Therefore, the Doppler effect formula can be written

$$\frac{f'}{f} = \frac{1}{\gamma(1+v/c)}$$

By using this, one can argue that the relativistic Doppler effect can be considered as the classical result for waves in a fixed medium (5.4), combined with time dilation. However, the comparison is a little artificial, because in the classical case the two observers are not equal: one is moving relative to the medium (such as air) conveying the waves, and one is not. To compare the relativistic case with the classical one, the speed of light has to be compared with the speed of sound relative to the *receiver* in the classical calculation—w for waves received by Alice, and w–v for waves received by Bob.

7.3 Golden nugget 3: addition of velocities

Suppose that in addition to reference frames S and S′, there is a third reference frame S″. This could be the rest frame of a third bead on our wire. Let the three beads all have different velocities, but placed so that they all meet at an event 0, and then at event A a signal is sent from the first bead towards the others: see Figure 7.2.

Let v be the velocity with which S′ is observed to move in the rest frame of S, then the relativistic Doppler effect formula (7.8) applies.

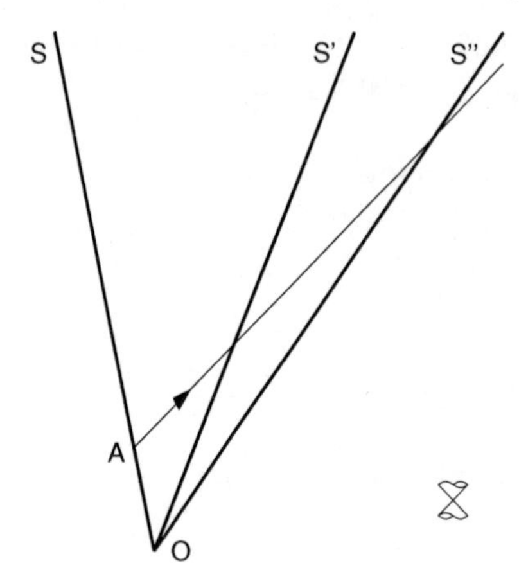

Figure 7.2: Spacetime diagram for three beads in relative motion along a line. At A a light pulse signal is sent from bead S towards the others. This enables us to find the relative velocity of S″ and S in terms of the other two relative velocities, by using the R factor three times (see text).

Now let us use the formula twice more. Define u to be the velocity with which S″ is observed to move in the rest frame of S′, then

$$\frac{f''}{f'} = \sqrt{\frac{1 - u/c}{1 + u/c}}$$

Inverting both sides gives

$$\frac{f'}{f''} = \sqrt{\frac{1 + u/c}{1 - u/c}} \tag{7.9}$$

Define w to be the velocity with which S″ is observed to move in the rest frame of S, then (using (7.8) again)

$$\frac{f''}{f} = \sqrt{\frac{1 - w/c}{1 + w/c}} \tag{7.10}$$

Now multiply (7.10) by (7.9) and compare with equation (7.8):

$$\frac{f'}{f''} \times \frac{f''}{f} = \frac{f'}{f} = \sqrt{\frac{(1 + u/c)}{(1 - u/c)} \frac{(1 - w/c)}{(1 + w/c)}} = \sqrt{\frac{1 - v/c}{1 + v/c}}$$

We can use this to obtain w as a function of u and v. After squaring both sides, multiply by $(1 + w/c)(1 - u/c)(1 + v/c)$ and rearrange, to obtain:

$$w = \frac{u + v}{1 + uv/c^2} \tag{7.11}$$

This equation is called the formula for 'relativistic addition of velocities'. Note that the answer from classical physics, '$w = u + v$', is not correct.

Equation (7.11) has some interesting properties. Notice that w always comes out less than or equal to c, as long as u and v are less than or equal to c. For example, if u is any speed, and $v = c$, then the formula produces $w = c$. This agrees with the Light Speed Postulate: it says that if a particle moving relative to me at speed u emits a photon, the latter travelling at speed c relative to the particle, then the photon also has speed c relative to me. This seems odd. It is one of the remarkable observations we first met in Chapter 3, but now that you understand something of Relativity it is not so odd if you think carefully about what we mean by relative velocity. In the formula, v is the speed that S$'$ has relative to S, *as observed by* S, u is the speed that S$''$ has relative to S$'$, *as observed by* S$'$, and w is the speed that S$''$ has relative to S, *as observed by* S. Observer *S* can reason like this: 'it's all very well you observer S$'$ reporting a speed u for S$''$ in your reference frame, but as far as I am concerned, your measuring sticks are all shortened, and your timing is too slow. After allowing for that, I get the formula (7.11).' The formula can indeed be derived from time dilation and space contraction, but the algebra is a little more involved; that method is typically described in university textbooks. Our derivation (learned from Hermann Bondi) is equally correct and complete. Figure 7.3 shows the result graphically.

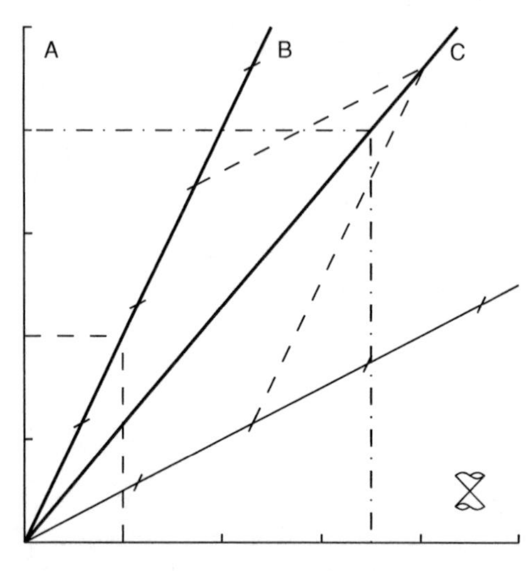

Figure 7.3: Addition of velocities in one dimension. The spacetime diagram shows three worldlines A, B, C, with tick marks indicating the passage of time in the rest frame of A and B (these are placed so as to allow correctly for time dilation, this will be clarified in Chapter 8). Position axes are also shown for the rest frames of A and B, with tick marks at unit steps of distance in those reference frames. Let us call the three particles Alice, Bob, and Charley. The dashed lines at the lower left indicate the speed of Bob relative to Alice. Bob takes two clock-ticks to travel one rod length, so the relative speed is $0.5c$. The sloping dashed lines can be used to read off the speed of Charley relative to Bob: it takes 3 clock-ticks (says Bob) for Charley to travel along two rod lengths of Bob's frame, so the relative speed is $(2/3)c$. The dash-dotted lines indicate the speed of Charley relative to Alice: in four clock-ticks (says Alice) Charley travels along 3.5 rod-lengths of Alice's frame, so the relative speed is $(7/8)c$. This result agrees with the formula given in equation (7.11). The diagram also helps to see why the relative speed does not emerge greater than c.

Equation (7.11) lets us explore the idea of jumping aboard ever faster rockets in an attempt to catch up with a pulse of light. We fire off a light pulse, and then jump aboard a rocket travelling at $0.8c$ (four fifths of the speed of light). So far we have achieved a speed of $0.8c$ relative to the launch pad on Earth. Now we enter a fast exploration vehicle, emitted forwards from the rocket at speed $0.8c$ relative to the rocket. Equation (7.11) says that our speed relative to the Earth launch pad is now $(0.8 + 0.8)c/(1 + 0.64) = 0.97561c$. Now climb aboard an escape pod fired forward from the exploration vehicle, at speed $0.8c$ relative to the latter. Equation (7.11) says that our speed relative to Earth is now $(0.8 + 0.97561)c/(1 + 0.780488) = 0.99726c$. Still not enough! In desperation, fire a probe forward from the escape pod, at speed $0.8c$ relative to the latter. Its speed relative to Earth is $(0.8 + 0.99726)c/(1 + 0.797808) = 0.9997c$. You get the idea: no matter how hard you try, giving yourself a boost from a fast-moving platform will never succeed in bringing your speed relative to anything else above the speed of light.

The whole attempt was utterly futile of course, because in the rest frame of the final probe, the relative speed of light pulse and probe is, according to (7.11), $(1 + 0.9997)c/(1 + 0.9997) = c$. In other words, the original light pulse is still disappearing into the distance at speed c, even for an observer sitting on the probe!

PUZZLE. The star-ship *Enterprise* is fleeing from a Klingon battle -cruiser. *Enterprise* zooms past a lonely asteroid at speed $0.9c$ relative to the asteroid. The Klingon battle-cruiser can only attain $0.5c$ relative to the asteroid. The *Enterprise*'s deflector shields can protect it from laser attack, but sensors indicate that the Klingons have unleashed their deadly proton torpedo, which can overwhelm the shields. The *Enterprise*'s database reports that such torpedoes are released at maximum speed $0.7c$ relative to the firing tube on the Klingon ship. Is the *Enterprise* safe?

CHALLENGE. A flea stands on the ground with a smaller flea on its back. This smaller flea carries a yet smaller one. If the fleas leap in sequence, the bottom one first, each achieving speed v relative to the surface it jumped off, then what is the final speed of the smallest flea relative to the ground?

In the case of three objects such as three beads moving along a line, there is another speed in which we might be interested: namely, the rate at which the distance between S and S$''$ increases, *as observed by* S$'$. That rate is given straightforwardly by $u + v$—the sum of two velocities both observed by S$'$. So why is w not equal to that? This requires some thought.

Note that $u + v$ is a speed, but it is not a relative speed. That is, it is not the speed of motion of any of the beads in the rest frame of another. Rather, it is the rate of change of a distance defined in reference frame S$'$—the distance between bead S and bead

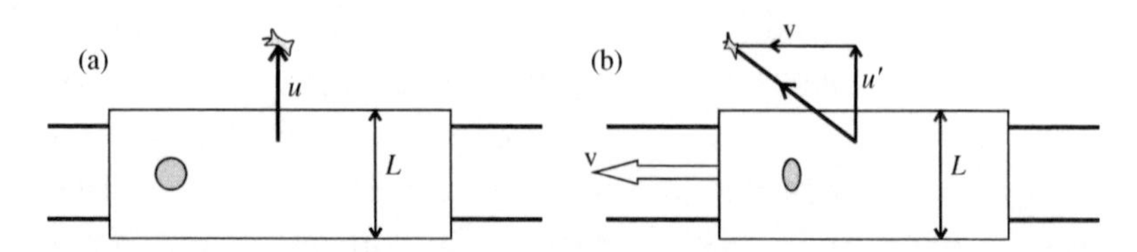

Figure 7.4: A passenger throws an apple core out of the window of a train. (a) Situation in rest frame of the carriage. (b) Situation in rest frame of the ground.

S$''$. But no physical object is moving at this speed. To convince yourself of this, imagine a spring attached to the two outer beads. Certainly the length of such a spring would be increasing at the rate $u + v$, when observed in reference frame S$'$, but no part of the spring is moving at that speed. When either of the other observers considers the spring, they find one of its ends to be at rest in their reference frame, and the other end to be moving at speed w given by (7.11).

To distinguish the two ideas, sometimes $u + v$ is called a 'closing speed' while w is a relative speed. A closing speed can be greater than the speed of light (up to a maximum $2c$), but this does not break the Light Speed Postulate because no signal is travelling at $u + v$. We shall discuss this point in more detail in Chapter 9.

7.3.1 TRANSVERSE VELOCITY

Now that we have a good grasp of the addition of velocities for objects moving along a line, let's take a look at another simple case: transverse velocity.

Imagine that a passenger on a train throws an apple core out of the window. Suppose that in the rest frame of the train, the apple core moves directly out, at speed u, on a trajectory at right angles to the train (see Figure 7.4). Then, to an observer on the ground the apple has both the forward velocity v of the train, and also a transverse velocity u'. In classical physics one would have $u' = u$, but it is not hard to see that u' will be smaller than u according to Special Relativity. For, suppose the apple core first flies across the carriage, of width L, before leaving through the window. In the reference frame of the train, it takes a time L/u to do that. In the reference frame of the ground, life on the train is going in 'slow motion' (time dilation), so the apple takes longer to cross the carriage. However, we have already established that there is no contraction of the carriage in the transverse direction. Therefore, the apple is taking a longer time to cover the same transverse distance, so its transverse velocity must be smaller by a factor γ.

To be certain that the factor is exactly γ we need to reason a little more closely. Instead of throwing an apple core out of the window (something which might be considered dangerous in any case), let the passenger conduct an experiment inside the carriage: she rolls a ball along a table top and it bounces off the side of the carriage wall. We suppose that in the carriage frame, the ball moves at constant speed u along a wooden tray of rest length L_0, at right angles to the wall, then bounces off the wall, and returns at speed u along the tray to its starting point. Then there are two events *at the same location in the carriage*—'ball leaves' and 'ball returns'—separated by time

$$\tau = 2L_0/u \tag{7.12}$$

in the carriage frame. This is a *proper time*, because it is a time interval between two events as observed in the frame where the events occur *at the same place*. That is why we used the symbol τ. This makes it straightforward to apply the time dilation formula: in any other reference frame the time interval between the same two events is longer by the time dilation factor.

Now consider those same two events as observed by someone standing on the ground. The events are separated by the time interval

$$t = \gamma\tau = \frac{1}{\sqrt{1 - v^2/c^2}}\tau \tag{7.13}$$

(time dilation), where v is the speed of the train relative to the ground. It is important to be clear that the speed entering in the formula for γ here is the relative speed of the reference frames, not the speed of the ball. In changing between reference frames there is no change in transverse length, so the wooden tray still has length L_0 in the ground reference frame, but now the ball travels down it and back in time t. Therefore its transverse speed is

$$u' = \frac{2L_0}{t} = \frac{2L_0}{\gamma\tau}$$

where we used (7.13) to get the second version. Combining this with (7.12) we obtain

$$u' = \frac{u}{\gamma_v} = u\sqrt{1 - v^2/c^2} \tag{7.14}$$

This is the conclusion: *transverse velocity is reduced by a factor γ_v, for a reference frame with relative speed v*. We put the subscript v on the γ symbol here as a reminder that this γ is the one associated with the speed v not u. This result will be useful in Chapter 10 when we consider momentum and energy and derive the most golden nugget of all: $E = mc^2$.

In the reference frame of the ground the ball has a forward as well as a transverse motion. Its total velocity is made of a component of size v in the direction of motion of the train, and a component of size u' at right angles to that. Therefore its total speed w, relative to the ground, is given by

$$w^2 = (u')^2 + v^2$$

(Pythagoras again!) Employing (7.14) we have

$$w^2 = u^2 + v^2 - \frac{u^2v^2}{c^2} \tag{7.15}$$

Now you can play with this formula, as we did with equation (7.11), to ensure that the principles of the theory are upheld. For example, if $u = c$ (imagine shining a torch out of the window) then we obtain $w = c$—the Light Speed Postulate again. If u and v are both less than c, then so is w, and so on.

Transverse velocity occurs in many everyday situations. Think of a bird flying directly towards you, for example. Its wing-tips move up and down, in a direction transverse to its flight path. According to Special Relativity their transverse motion is slower in your reference frame than in the reference frame of the bird. Of course, the speeds of birds are small, but the same effect is readily measured in atomic physics, when atoms move fast while vibrating internally, and it is observed every day in high-energy particle collision experiments, and in astronomy.

Puzzle. Does the Global Position System (GPS) need to take time dilation into account? (Imagine the newspaper headline: 'Army platoon lost in desert: "we did not understand Relativity" admits Chief of Staff.') Here is some data to help you assess this: the satellites of the system each orbit twice a day at an orbital radius of 26, 600 km. The essence of the system is that a receiver picks up signals from four or more satellites, and uses timing information in the signals to calculate the time of flight of electromagnetic waves from the satellite to the receiver. The satellites also report their own position very accurately. This allows the receiver to infer its position.

You could try to figure this out for yourself; but to be a little more friendly, here is a guide. The idea is that the satellites need to have reliable clocks on board; but they are not reliable enough to be left to themselves, so they are checked and corrected regularly by timing information sent from control stations on the ground. The problem is that it is not feasible to do this more than a few times per day. So the issue is, will time dilation cause too large a timing error to build up, if it is not taken into account in the design of the system?

First, use the information about the orbits to find the speed of the satellites. You need to recall that the circumference of a circle is about 3.14 (π) times the diameter, so roughly six times the radius. You should find $v \simeq 3.9$ km/s. Use this to determine the γ factor: it is very close to 1, so it is best to find out by how much it is larger than 1 by using equation (6.7). You should find the answer is that γ exceeds 1 by approximately one part in ten billion (10^{-10}). Next, what size of timing error in the satellite could result in the receiver reporting its position incorrectly by 100 metres? (You only need to use the known speed of light to estimate this). How long will it take a clock on a satellite to build up this amount of error, if no allowance is made for time dilation compared to a clock on the ground? Is it feasible to update the satellites with timing information from ground-based clocks often enough to avoid this source of error—or must the relativistic correction be built into the system?

(In fact, clocks in orbit require two relativistic corrections: the time dilation owing to their motion, and a gravitational effect. Here we have just examined the motional effect.)

8

Navigating in spacetime

In this chapter we will consider some famous paradoxes of Special Relativity: that is, things which appear paradoxical at first sight, but which when understood are no more paradoxical than $1 + 1 = 2$. Sometimes we can learn a great deal from a well-chosen paradox, and this will be the case here.

Before we go further let us pause for a moment, because I hope you may be asking two simple questions: 'What causes time dilation?' and 'What causes space contraction?'

The answer to both questions is 'This is a bit like asking, "what causes a circle to be round?"' We do not need to look for a cause of a circle being round, because that is merely a defining property of a circle. In a similar way, time dilation is a fundamental property of what we call time, and its 'cause', if you want to put it that way, lies in the nature of spacetime. We have a spacetime in which all inertial reference frames are equivalent, and in which there is a maximum speed for signals. In such a spacetime, dynamical evolution, when considered in different reference frames, will be found to obey the time dilation formula, and the lengths of objects will be found to obey the space contraction formula.

Notice that effects such as time dilation and space contraction are not merely a matter of words and definitions. They lead to testable physical results, such as the arrival of muons on planet Earth. They are confirmed every day in experiments involving high-velocity particles, and in astronomical observations of high-velocity stars and galaxies. Even without those confirmations, the theory is logically sound and beautiful—so beautiful that once we appreciate it we begin to long for it to be true, and classical Newtonian physics begins to seem the strange and awkward partner. For classical physics allows something deeply suspicious to a good physicist: namely, an unlimited speed for signals. It just does not seem right to our physical intuition that an impulse could travel across the universe in the twinkling of an eye. But as soon as we insist on a finite maximum, suddenly Special Relativity comes into being 'before our eyes', as it were, and the universe makes better sense.

8.1 The pole and barn paradox

Here is the first lovely paradox of this chapter.

Suppose an athlete carries a horizontal pole of rest length 4 metres, and runs fast towards a barn, also of rest length 4 metres, at the (admittedly audacious) speed of $0.866\,c$. In the rest frame of the barn, the pole is contracted by the factor $\gamma \simeq 2$, so its length is 2 metres and it will easily fit inside the barn (see Figure 8.1a). Indeed, as soon as the back end of the pole enters the barn, the barn door is closed so that the pole is totally enclosed inside the barn. Shortly after, when the front of the pole reaches the far end of the barn, a back door there is opened and the pole and athlete emerge.

But now consider the same events from the point of view of the athlete (Figure 8.1b). He finds himself to be carrying a 4-metre pole, with the barn fast approaching him. According to Special Relativity, it is now the barn that suffers a contraction to 2 metres length, so the pole cannot possibly fit inside it. The barn doors cannot both be closed with the pole inside the barn. It is impossible!

'This is pure nonsense, a clear contradiction,' says a student, 'it disproves Lorentz contraction and therefore Special Relativity.'

What is going on? Is the student right? Does the pole fit inside the barn, or not? You are encouraged to try to resolve this yourself before reading on. Here is a strong hint: it turns on the relativity of simultaneity.

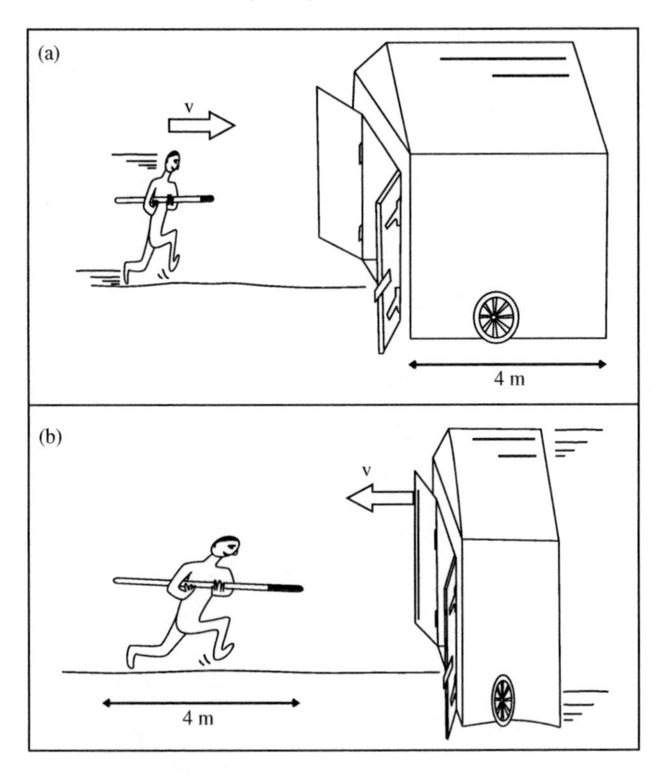

Figure 8.1: The pole and the barn. (a) An athlete runs fast towards a barn, carrying a pole of rest length 4 metres. The barn is also of rest length 4 metres, but owing to Lorentz contraction the pole is only 2 metres long, so it will comfortably fit inside the barn, with all the barn doors closed. (b) shows the *same* athlete and barn, but in the 'rest frame' of the athlete. He looks stouter and still needs to run, because the ground is whizzing by under his feet as the barn approaches. From this point of view, it is the barn that moves and thus is contracted, while the pole has its full rest length. The pole cannot possibly fit inside the barn! Is this a contradiction? What is going on?

Resolution. The spacetime diagram shown in Figure 8.2 explains what happens. On the diagram we draw four worldlines—those of the tip and tail of the pole, and of the front and back of the barn. We shade in the regions between these lines, to represent all the particles of the pole (light grey) and the interior of the barn (dark grey). Event B is the closing of the front door as the tail of the pole passes inside the barn, and event C is the opening

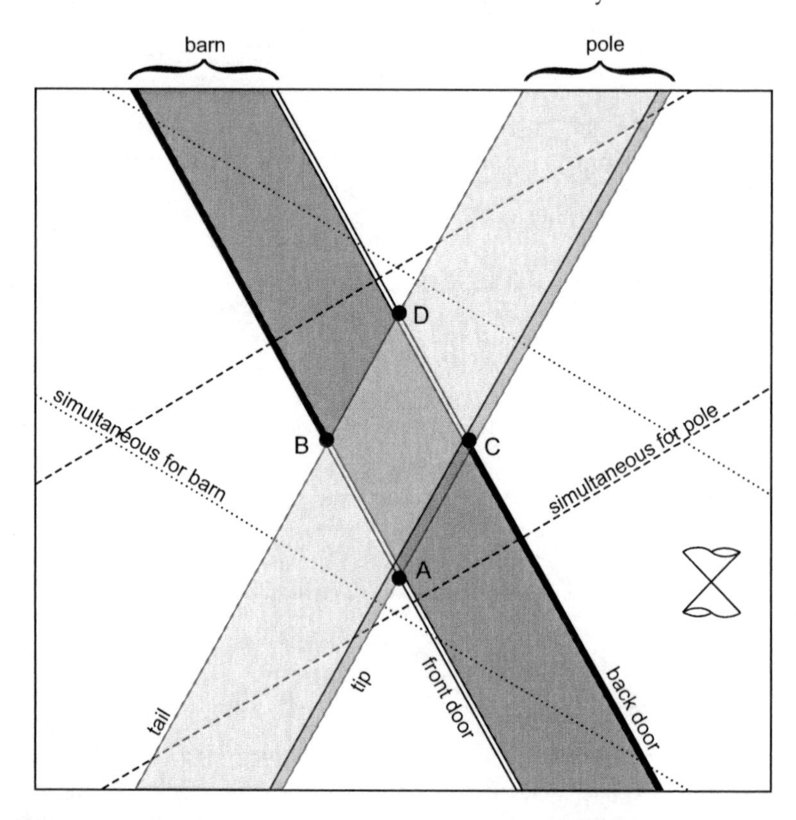

Figure 8.2: The pole and barn paradox resolved. The figure is a spacetime diagram—essentially the same one as Figure 6.18, but here reproduced with more details. We only concern ourselves with the extent of the pole and the barn along the line of relative motion, in order to have a problem in one spatial dimension only, and therefore a two-dimensional spacetime diagram. The worldsheet of the pole is shaded light grey, that of the barn dark grey. Event A is the tip of the pole entering the barn, event B is the tail of the pole entering the barn and the front barn door closing, event C is the back door of the barn opening as tip of the pole exits, and event D is the tail of the pole leaving the barn. The dashed lines are example lines of simultaneity for the rest frame of the pole, and the dotted lines are example lines of simultaneity for the rest frame of the barn. In both reference frames A precedes B and C which precede D. In the barn's rest frame B happens before C, so the pole is for a while wholly enclosed within the barn, with both doors closed. In the pole's rest frame, C happens before B, so the pole is never wholly enclosed within the barn. Instead after C it pokes out of both (open) doors for a while, until at B its back end enters the barn. It subsequently leaves and continues on its way.

of the back door as the tip of the pole reaches the far end of the barn. To be sure that the shaded 'worldsheets' show two objects of the same rest length, we select two equal and opposite slopes for their worldlines on the diagram, and draw them symmetrically. To aid in interpreting the diagram, two lines of simultaneity are included for each of the two rest frames, at the bottom and top of the diagram.

Use the 'moving slot' method to read the diagram. That is, first align your slot or 'time-slice' with the line of simultaneity of the rest frame of the barn. You should see that the region occupied by the barn, along this line of simultaneity, is indeed twice as long as the region occupied by the pole. Now slide the time-slice slot upwards, being careful to maintain it always parallel to the lines of simultaneity of the barn. You should find that event A happens first, then event B, after which for a short while the pole is completely inside the barn, and then event C happens, and the pole begins to emerge, leaving fully at event D.

Now, *using the same diagram*, orient your slot or 'time-slice' with the line of simultaneity of the rest frame of the pole. Along this line of simultaneity, the barn is now half the length of the pole, in agreement with the prediction of space contraction. Slide your time-slice upwards again, but now maintaining it parallel to the lines of simultaneity of the pole. You should find that the paradox is resolved because the time ordering of events B and C is now reversed. Event C—the tip of the pole leaving the barn—happens before the tail has reached the barn. For a while the pole pokes out through both ends of the barn. Then event B happens (the closing of the front door), and subsequently event D. In this frame of reference, the pole is indeed never wholly inside the barn, but there is no contradiction: there is simply a single well-defined set of events in spacetime. The pole and barn paradox is thus beautifully resolved.

Let us pause here for a moment, to savour the wonderful elegance of this resolution.

. . . (Some time passes) . . .

Now for the sting in the tail. Suppose the back door of the barn is never opened to let the pole out, but instead a very strong concrete block is placed there. The front door is still closed once the back end of the pole passes it. The barn still has rest length 4 metres, but suppose the pole's rest length is 6 metres (so the contracted length is 3 metres). Can such a pole be stopped inside the barn?

In this case, in the rest frame of the barn, the pole enters the barn cleanly without collision, because its Lorentz-contracted 3-metre length fits comfortably inside the barn. Next, the front of the pole collides with the block. The pole will now be compressed, and eventually all of it will stop. Assuming it survives (perhaps it is made of rubber) then it must either bend or finish in a highly compressed non-equilibrium state, or it must burst back out of the door.

In the rest frame of the pole, the front end of the pole collides with the concrete block well before the back end of the pole reaches the door, so the question arises, *does the back end of the pole still enter the barn?*

A spacetime diagram makes it obvious that indeed it does (try drawing one), and the interesting point is that after the tip of the pole hits the concrete, the back of the pole continues its *constant-velocity* motion for a while, completely oblivious of the crunching collision in progress at the front! The back end moves cleanly into the barn, and only somewhat after entering does it receive any hint that a collision is underway at the front. Indeed, the event of entry into the barn of the back end of the pole is right outside the future light cone of the first collision event at the front end, so it cannot be influenced by the initial collision. The speed of the back end of the pole only starts to fall when a wave of compression travelling down the pole reaches it. Such compression waves typically travel at around the speed of sound.

When the compression wave reaches the back of the pole, the latter begins to decelerate. Since in this scenario the pole is brought to rest while being prevented from regaining its natural length of 6 metres, it finishes in a compressed state, with forces acting on its two ends. Note that this type of compression is different from Lorentz contraction.

8.2 The bug and the rivet

The last part of the pole and barn paradox is very closely related to a paradox which is presented in Figure 8.3. A bug crouches in the bottom of a 10-mm deep hole in a table-top. A rivet of rest length 8 mm is going to be placed in the hole. Since the rivet is not long enough to reach to the bottom of the hole, this could be done without squashing the bug.

Now suppose the rivet approaches the hole at high speed, such that $\gamma = 2$ for the relative motion. First consider the situation in the rest frame of the bug: see Figure 8.3(b). The hole is not moving relative to the bug, so it is uncontracted and remains 10 mm deep. The rivet, on the other hand, is moving, so it is Lorentz contracted to a length 4 mm. The bug feels safe in his hole because he thinks the fast-moving rivet is much too short to reach to the bottom of the hole and squash him.

However, the *same* scenario looks different from the perspective of the rest frame of the rivet: see Figure 8.3(c). Now the hole is contracted to 5 mm, while the rivet is 8 mm long. It can easily reach the bottom of the hole and will surely squash the bug. So what was wrong with the bug's reasoning?

To resolve this paradox it is helpful to keep fixed in one's mind two things. First, the fate of the bug is not frame-dependent. There is just one bug, and if it dies then there is a death-event which all observers will observe; they differ only on its timing and distance relative to other events. Second, whichever reference frame provides the clearest or simplest description is the one to trust in the first instance, if you are not sure. In this example, figure 8.3(c) makes it very clear that the bug must die: in this frame (the rest frame of the rivet), up until the moment when the end of the rivet reaches the bottom of the hole everything is extremely simple—just two things moving together at constant velocity.

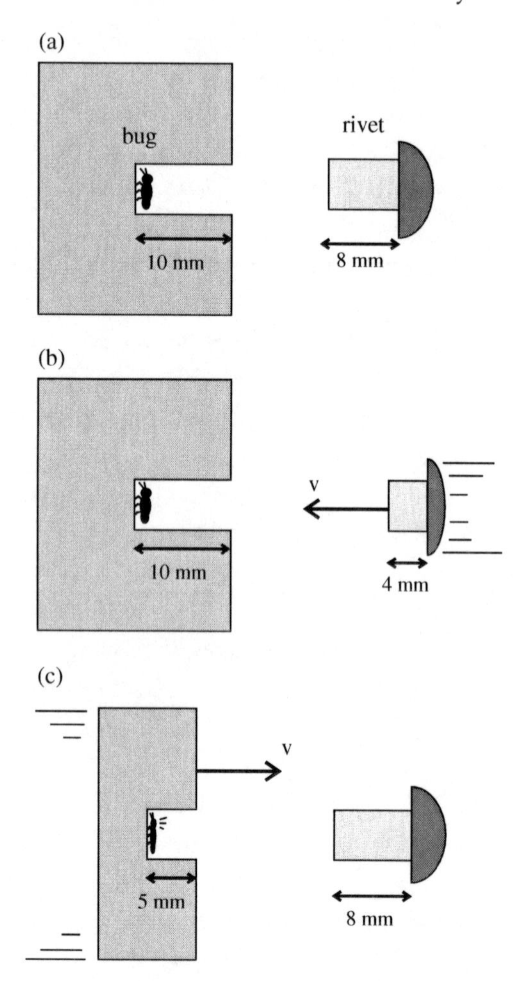

Figure 8.3: The bug and the rivet. (a) A bug sits at the bottom of the 10-mm deep hole. If the rivet, of rest length 8 mm, is placed in the hole, it is not long enough to reach the bottom, so the bug will not be squashed. (b) If the rivet approaches the hole at high speed, then in the rest frame of the bug the rivet is Lorentz-contracted, so it is shorter still, and the bug feels safe. (c) The *same* situation, but as observed in the rest frame of the rivet. Now things do not look good for the bug: his hole is Lorentz-contracted to only 5 mm deep, so surely the 8-mm-long rivet will easily squash him. Which is correct? Does the bug survive?

What the poor bug failed to understand is that when the head of the rivet catches onto the table surface at the top of the hole, the bottom of the rivet does not immediately stop moving relative to him. The information that there is something impeding the motion takes a while to reach the bottom of the rivet, so in the bug's rest frame the rivet begins to stretch. It stretches all the way to the bottom of the hole (as the picture (c) insists that it must) and flattens the bug, before eventually shrinking back to its ordinary length again. The objective facts are summarized by the spacetime diagram shown in Figure 8.4.

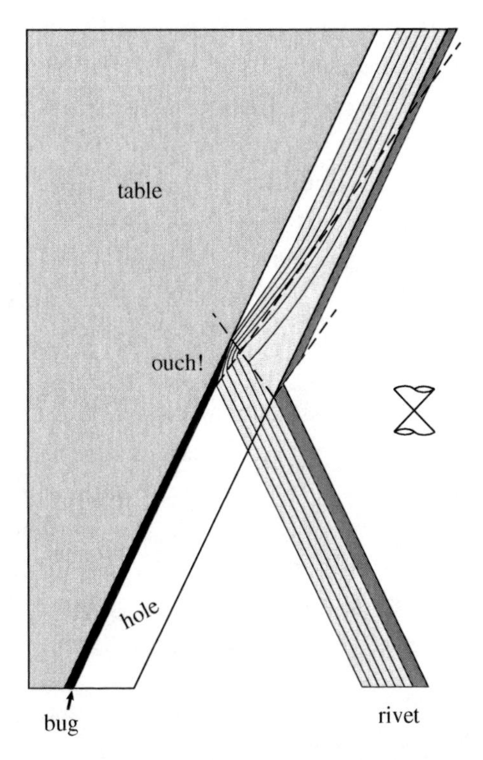

Figure 8.4: Spacetime diagram for the bug and the rivet. The lines in the rivet show worldlines of example particles. The dashed lines show the propagation of compression or extension waves in the rivet. The particles begin to change their motion when such a wave reaches them. There is no reason for the end of the rivet to stop approaching the bug before it hits the bug.

8.3 Calibrating spacetime diagrams

In order to make further use of spacetime diagrams, we will now establish how they are calibrated.

We always orient the diagram such that photon (or light pulse) worldlines going each way have equal slopes on the page. It is also convenient to choose scales such that these slopes are at 45°.

Recall Section 6.2 on simultaneity, and Figure 6.13 on p.95. There we showed how to identify sets of simultaneous events in a given reference frame, and sets of events at the same position in the given reference frame. We did not then establish the scales; that is, how much distance on the paper ought to correspond to a given interval of time or space. The overall scale of the diagram does not matter of course, but we need the scales of different reference frames to be mutually in the right proportions, so that they exhibit the correct amount of space contraction and time dilation relative to one another.

In classical spacetime diagrams this was easy: there, a given vertical distance on the paper corresponded to a given interval of time, for all reference frames, and a spatial distance in any given reference frame was obtained by a horizontal distance on the paper.

In special relativistic spacetime diagrams, there is no universal direction for a time or space axis. Instead, the worldline of each uniformly moving particle acts as its own 'time axis'. Thus we know where to draw the time axis for any reference frame. We also established how to identify a line of simultaneity for any reference frame. The line of simultaneity passing through some chosen 'zero' event on the time axis can serve as the distance axis or 'x axis' for that reference frame.

The diagram appears to single out one reference frame for special consideration: namely, the one whose time axis and distance axis are orthogonal on the diagram. This is a weakness of the diagram, because it hides the Principle of Relativity which in fact spacetime obeys, but if we use the diagram correctly then we will not obtain from it any statements that are counter to the Principle. We have already encountered the same weakness of classical spacetime diagrams in Chapter 4.

It only remains for us to allocate tick marks to the time and distance axes of all reference frames. But be careful: a given vertical distance on the paper does not correspond to the same time interval for all reference frames, and neither does a given distance along each time axis. Instead we must respect the time dilation formula (6.5). To do this, first mark an equally-spaced sequence of ticks on a vertical line. This vertical line is the time axis of the reference frame (t, x) whose time axis is vertical on the diagram (we have to start with some reference frame, so we may as well start with this one). The ticks indicate unit time steps in this reference frame. Next draw some other worldline, intercepting the first one at one of the ticks. This worldline is the time axis of reference frame (t', x'). Let the crossing point serve as the zero of time for both reference frames. We want to find the event on the new worldline at $t' = 1$, $x' = 0$. Note that t' is the *proper time* between the crossing of the wordlines and this 'tick' event. Therefore, it occurs at the dilated time $t = \gamma t'$ in the first reference frame. This is sufficient to enable us to locate it: see figure 8.5. (You may need to recall the definition of proper time on p.133.)

We have just established that for a general reference frame S′ of speed v relative to the first frame S, the event of the first clock tick in S′ occurs at $t = \gamma$, $x = vt$ in S. By eliminating v from this pair of equations, we obtain the relationship between t and x for all such events: square the first equation $t = \gamma$ to find $t^2(1 - v^2/c^2) = 1$, then replace vt by x, to obtain

$$t^2 - \frac{x^2}{c^2} = 1 \tag{8.1}$$

The set of all events (t, x) that satisfy this equation fall on a smooth cup-shaped curve on the spacetime diagram: see Figure 8.6. This sort of curve is called a 'hyperbola'; the same shape can be obtained by cutting a cone-shaped object in two, at an oblique angle, and looking at the outline of the resulting cross-section.

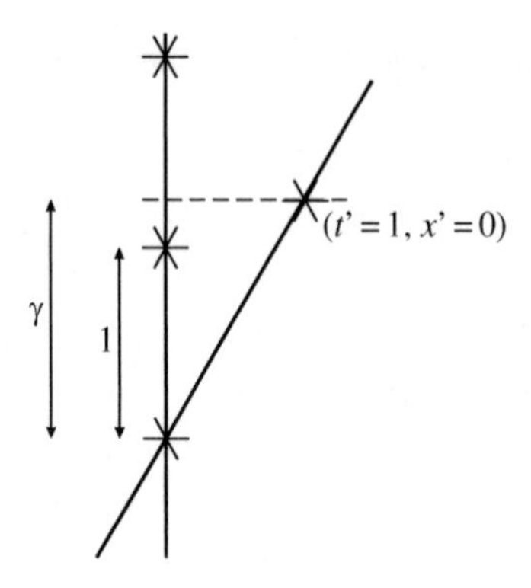

Figure 8.5: Calibrating a spacetime diagram. Two worldlines are shown, representing two particles with relative speed v. If the first worldline is vertical then v is indicated by the slope of the second worldline (for this example it is 0.6 c). The $*$ marks on the vertical worldline indicate the timescale for one particle, which can be chosen arbitrarily. Once this is done there is only one choice for the timescale on the other worldline: it must correctly indicate time dilation. Therefore the event shown at (t'=1, x'=0) must occur at time $t = \gamma$, as shown. For this example, $\gamma = 1/\sqrt{1 - v^2/c^2} = 5/4$.

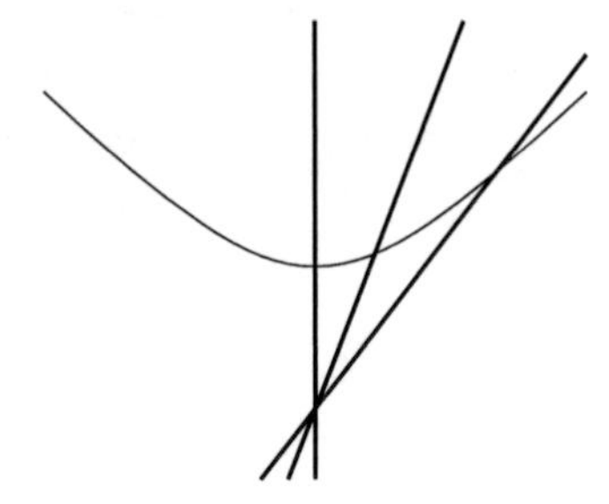

Figure 8.6: Calibrating a spacetime diagram, part 2. Three worldlines are shown, for three speeds relative to the vertical worldline (the speeds are 0, 0.4 c, and 0.8 c in this example). Each straight worldline can be regarded as the time axis for the corresponding reference frame. The curve is the hyperbola described by equation (8.1), which allows the timing in every reference frame to be calibrated correctly on the diagram. We suppose clocks are synchronized to read zero at the crossing point of the worldlines. The hyperbola is drawn so that the point on the diagram where it intersects the vertical axis is at 1 time unit (1 second) after zero, for the reference frame whose time axis is shown vertically. Then, the event where the hyperbola intersects the time axis of any other reference frame is the clock tick event at 1 second after zero, *for that reference frame*. Thus we can quickly calibrate any set of reference frames. It is not surprising that the clock tick events are more spread out in the horizontal direction on the diagram when the corresponding clock has a sloping worldline, but we see that they are also more spread out in the vertical direction. This is the time dilation effect.

The curve shows that the size of a 'unit step in time' appears larger and larger on the diagram, as we consider reference frames whose time axis is sloped more and more away from the vertical. Such a 'unit step in time' is a time-like interval of unit size, so in a sense it is a unit step in spacetime.

Perhaps we would find it easier if the diagram were to succeed in showing a unit step in spacetime as a step of given size on the paper, no matter in what spacetime direction. It does not come out that way, however. This is because time is not quite the same as space, so the geometry of spacetime is not exactly the same as the geometry of a piece of paper, and we have to live with this. The spacetime diagram is a powerful aid to understanding spacetime, but the diagram is not itself spacetime (it occupies a region of space, and it extends through time . . . until the paper is eventually destroyed).

Having marked the ticks on the time axes, the ticks of unit distance on the distance axes can be entered easily, so as to ensure photon worldlines have the same speed in all reference frames. This is satisfied if the distance-axis ticks of any given reference frame have the same separation along their axis as the time ticks do along their axis.

The illustrative diagram in Figure 6.13 already had all of these considerations correctly incorporated.

8.4 The twin paradox

The most celebrated paradox in Special Relativity is the twin paradox. It provides a powerful example of a relativistically predicted phenomenon, and in the course of resolving it we will make an important discovery.

Consider two twins. We want to consider twins because they have the same date of birth, so we know they are the same age (as long as they have not been travelling, as we shall see). One of the twins is named Adam, after the Earth, and the other is named Astra. Astra has a longing to go space travelling, while Adam prefers to stay at home. They part company one day at the rocket launch pad. Astra blasts off into space, and after accelerating for some months reaches a very high velocity away from Earth—a large fraction of the speed of light. She coasts at constant velocity for a few years, enjoying the view through the window, then fires her rockets again to slow and reverse her motion, then coasts back towards Earth, again at high speed relative to the Earth. Upon reaching Earth she touches down and enjoys a happy reunion with her twin brother.

Observed from the rest frame of Earth, Astra has been moving at high velocity; therefore, she and everything in the rocket undergoes time dilation. This means that during the few years of rocket time while she is moving away, several decades pass on Earth. On the return journey the same happens again: only a few years pass in the rocket, while many decades pass on Earth. Therefore, when the twins meet again, Astra has aged by just 6 years, but Adam and the rest of the population of Earth have aged by 60 years.

So far so good. This is an interesting prediction, but not yet paradoxical in the sense of a paradox of logic. However, consider the point of view of Astra. As far as she is concerned, it is the Earth that has moved away at high velocity, and then returned at high velocity. This implies that the whole situation is symmetrical: Adam and Astra have undergone equivalent motions, so they cannot possibly have different ages. According to this argument, time dilation, and therefore Special Relativity, is self-contradictory and wrong.

Note that we have a clear contradiction: at the meeting, either the twins have the same age, or not. The ages cannot be both equal and unequal. Their aging would be indicated by all sorts of things: their physique and memories, their watches, all the keepsakes such as photograph albums and video footage they gathered, the journals they wrote, and so on.

Again, before reading on, you are invited to try to resolve this paradox yourself. If you do not want to see the resolution, look away now! Here it comes!

Resolution. The paradox turns on the claim that Adam and Astra had symmetric experiences: for each of them, the other one moved away and came back. If their experiences were really the same, then their net ageing must be the same. However, this overlooks a crucial fact: Astra experienced three accelerations, while Adam did not. Astra experienced acceleration when her rocket motors fired at the start, and during her change of direction in order to make the return journey, and when she landed. No rocket motors were attached to the Earth: apart from the modest effects of its orbital motion, it remained an inertial reference frame throughout (and furthermore, we introduced planet Earth only for the sake of story-telling: we could imagine Adam living on a space-station instead, floating free without orbiting any planet or star). Acceleration is crucial because Astra can tell she was accelerated without needing to 'look outside' the rocket: it is not merely a relative property of her motion with respect to Adam. Many physical phenomena could serve as an accelerometer which she could use to check this. Equally, Adam can be sure that he was not accelerated. Therefore their experiences are not symmetric, and there is no logical requirement that their ageing should be equal.

Drawing the worldlines on a spacetime diagram (Figure 8.7) makes it blindingly obvious that the situation is not symmetric: Astra's worldline is bent, while Adam's is straight!

A period of acceleration is important because it changes the *shape* of the worldline, and it can do this without necessarily taking up a large fraction of the whole worldline. This is similar to the fact that a bend in a road can have a large influence on the total length of a journey between two places, even though the bend itself is only a small part of the road: see Figure 8.8.

This completes the resolution of the paradox. In short, there is no paradox. The physical scenario remains an interesting one, however, so next we will look into it in detail, in order to find out precisely what Relativity predicts and what it can teach us about spacetime. It will teach us quite a lot!

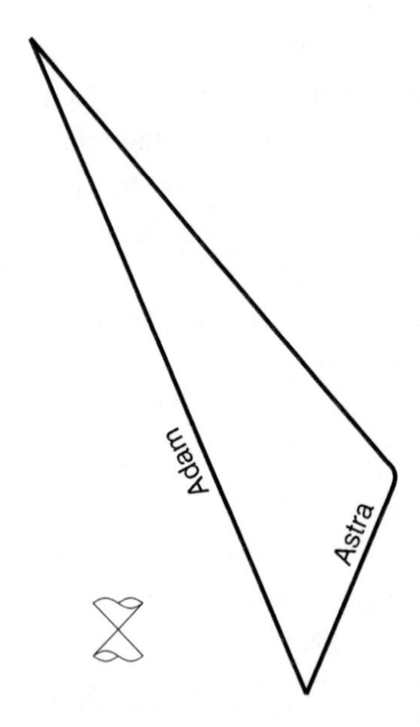

Figure 8.7: Spacetime diagram for the twin paradox. The diagram makes it obvious that the situation is not symmetrical between Adam and Astra, so really there is no paradox. There is nothing wrong with the relativistic prediction that Astra ages less than Adam.

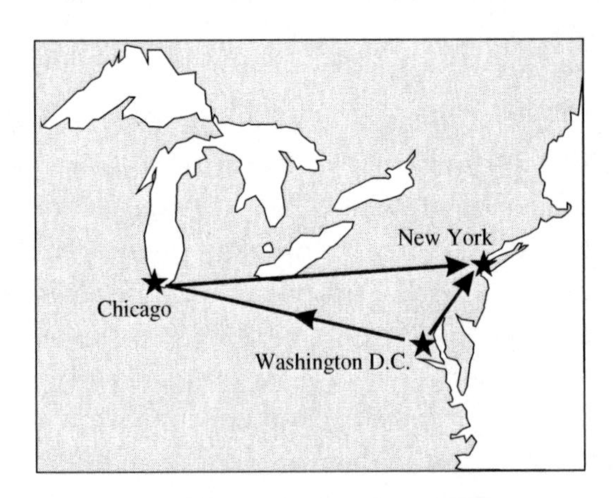

Figure 8.8: A road journey from Washington to New York. Two routes are shown: one a direct route, the other via Chicago. Someone might be puzzled as to why the route via Chicago is so much longer. After all, in both cases the car travels in a straight line almost all the time—the corner is only a small part of the total journey. Nevertheless, that seemingly innocuous change of direction is enough to make the total distance much longer.

First we will present the calculation to back up the numbers given in the story. The parts of Astra's journey that are at constant velocity are easy to assess, so let us first consider the accelerated parts.

We know from our everyday life that modest accelerations do not greatly affect time and ageing, so we can guess that time proceeds for Astra during her accelerations in a reasonably straightforward way. For example, it does not suddenly rush by so as to produce the extra 54 years of ageing that would be needed to make her finish the same age as Adam. Let us suppose that at the start, Astra fires her rocket engine for a continuous period of 11 Earth months, and that it provides an acceleration of 98 metres per second per second. This is $10\,g$, where g is Earth's gravity. I chose this number because it is known that accelerations of up to $12\,g$ can be tolerated without undue discomfort or visual disorders, for someone lying down. It would take some special equipment to allow Astra to survive a 10-g acceleration for weeks on end, but certainly such a modest acceleration will not damage any electronic timekeeping devices on the rocket. They function perfectly normally even under much higher accelerations.

With these choices we can use equation (2.6) to find Astra's final coasting speed: it is $v = 0.995\,c$. The gamma factor for this speed is $\gamma = 10.0125$, and you can check that it increased roughly uniformly with Earth time, suggesting that time dilation during the acceleration period is on average a factor 5, so we estimate Astra aged around 2 months during the initial acceleration. A similar calculation applies to the period when she changes direction, and when she lands on Earth.

It turns out that these periods of acceleration are mere details, because they occupy only a small fraction of the total time. Most of the time Astra is coasting at $0.995\,c$ relative to the Earth, with $\gamma \simeq 10$, so during her outward journey taking three of her years, 30 years pass on Earth, and similarly for her return journey. Thus Special Relativity predicts that Astra ages by about six years in total, while Adam ages by 60 years.

Some people like to discuss the twin paradox while trying to avoid the acceleration part; for example, by considering three clocks, all in uniform motion, that are synchronized as they pass each other. One clock stays on Earth, the second flies alongside the rocket on its outward journey, and the third accompanies the rocket on its return journey. These three clocks have three straight worldlines, so we already know how to calculate the prediction of Special Relativity for them. This can help to emphasize that if one tries to argue that Astra will age 60 years after all, then one has a hard job, because something very bizarre would have to happen during her periods of acceleration. However, there is no need to be 'shy' of the acceleration part and try to avoid it: Special Relativity can handle it perfectly accurately.

The twin paradox is striking because it makes a very definite prediction about a very straightforward result: Adam has 60 years'-worth of experience, journals, and photographs, and Astra has 6 years'-worth of experience, journals, and photographs. This is not a statement about people a long way away from one another, or one in a rocket and one in a space-station: at their reunion, Adam and Astra are standing right next to one another in an ordinary way on *terra firma*.

This underlines once again that special relativistic effects are not merely about abstractions and definitions: they are about concrete realities. The twin paradox is observed in a myriad ways in physics experiments. For example, every time a particle travels around the ring of the great accelerator in CERN, Geneva, its behaviour is time-dilated compared to the 'stay at home' particles which make up the material of the vacuum chamber and associated detectors. However, this may seem far removed from everyday life. Because the twin paradox is so memorable, it was tested in a more familiar setting by J. C. Hafele and R. E. Keating in 1971. They flew four caesium atomic beam clocks around the world twice, on board regularly scheduled commercial jet flights. At the speed of a jet airliner the time dilation is a tiny effect, but one that is just observable using atomic clocks. After taking gravitation (General Relativity) as well as Special Relativity into account, the theory predicted that the flying clocks, compared with reference clocks at the US Naval Observatory, should have lost 40 nanoseconds during the eastward trip around the world, and should have gained 270 nanoseconds during the westward trip, with a 20-nanosecond uncertainty in both predictions, owing to the lack of precision in the flight information. In the event the clocks were observed to lose 59 nanoseconds during the eastward flight and gain 273 during the westward flight, with a 10-nanosecond imprecision in both numbers. Thus the experiment agreed with the theory to well within the expected precision, and confirmed it to within a few percent of accuracy. Other tests of Relativity have been much more precise, but this was one of few that examined the subject in the humdrum and messy situations of everyday life.

Special Relativity and acceleration

There is a strange myth that has grown up in the physics student community over many years, to the effect that Special Relativity cannot handle accelerations, and you need General Relativity for that. This is simply untrue, so I hope you will ignore it. Special Relativity can both correctly predict the trajectories of accelerated bodies in any inertial reference frame, and it can also correctly handle the internal evolution of such accelerated bodies. The myth has come from a muddled attempt to make a valid point about General Relativity. This is that in General Relativity one uses a higher level of abstraction in the mathematics, so that a single equation can describe motion in any sort of reference frame, including non-inertial frames. This is crucial for understanding gravity. The limits of Special Relativity arise in two ways. The primary limitation is that Special Relativity assumes that spacetime is 'flat', in a technical sense, while General Relativity handles a spacetime that can be warped or 'curved'. See an introductory text on this subject for more information.

Another limitation is concerned with accelerated reference frames; that is, not just how accelerated bodies behave, but what spacetime 'looks like' from the point of view of an accelerated observer. Special Relativity can handle the case of a reference frame undergoing constant acceleration very nicely. However, for completely general reference frames, in any state of motion, the mathematical techniques of General Relativity are required.

The conclusion is that Special Relativity can correctly treat any motion, accelerated or not, in a flat spacetime. It would normally be applied by choosing one or more inertial frames of reference and describing what goes on in them. To analyse motion in a flat spacetime from the perspective of a reference frame undergoing changing acceleration would require mathematical techniques 'borrowed' from General Relativity, so this is an intermediate case. A non-flat spacetime cannot be handled by Special Relativity because the zeroth Postulate would be invalid.

8.4.1 SIGNALLING IN THE TWIN PARADOX

To fill out the story of Adam and Astra, suppose now that the twins communicate with one another during the flight. They can perfectly easily do this by transmitting light pulses or radio signals. This way they can keep track of each other's progress—Astra monitoring events on Earth, and Adam watching to see how his sister is getting on. Figure 8.9 shows the propagation of the signals in spacetime. To reduce clutter on the figure, a more modest final speed for Astra is shown: $v = (3/5)c$, giving $\gamma = 5/4$. In the example shown, while 10 years pass on Earth, 8 years pass on the rocket. The acceleration periods are short, and we neglect their contribution to Astra's proper time. Each twin sends out a Christmas greeting once a year. These are received at a time separation given by the Doppler shift formula, equation (7.5), which gives $R = 2$. So on her outward journey, Astra receives Adam's Christmas message once every two years, and on her return journey once every half year (Figure 8.9a). The two parts of her journey last four years each for her, so she receives two 'slow' messages and eight 'quick' messages, and therefore returns fully aware of all Adam's experiences and noticing that he has had ten years'-worth of them. Adam meanwhile receives four 'slow' messages from Astra during his first eight years, and four 'quick' ones during his last two years (Figure 8.9a). Thus, while he aged

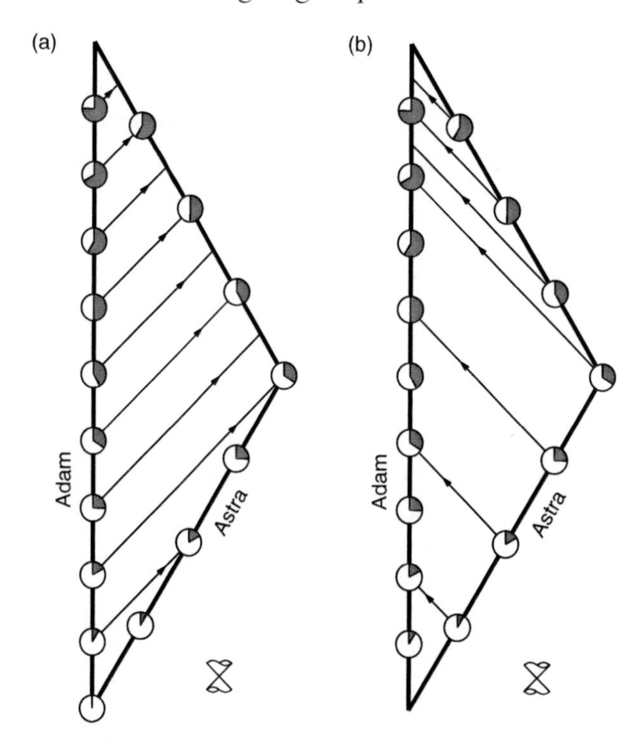

Figure 8.9: Spacetime diagram for the twin paradox, showing signals propagating between the twins. The periods of acceleration are assumed to be short enough to contribute negligible amounts of proper time for the rocket twin (Astra). The clock faces on the worldlines indicate yearly intervals of proper time. The left diagram (a) shows light-speed messages sent from Adam to Astra at the end of each of his years. The right diagram (b) shows light-speed messages sent from Astra to Adam at the end of each of her years. Astra receives two messages at a spacing once every two years, and seven at a spacing once every half year, plus a final message when she meets Adam, so she agrees that Adam aged ten years to her eight. Adam receives four messages at a spacing once every two years, and three at a spacing once every half year, plus a final message when they meet, so he agrees that Astra aged eight years to his ten years.

ten years, he notices that Astra has had eight years'-worth of experiences to tell him about.

If these signals were instead conveyed by a continuous video stream, then each twin sees a video image of the other moving about, either in 'slow motion picture' or 'fast motion picture', but they see differing proportions of the two types of video stream:

Astra sees most of Adam's experiences in 'fast motion picture', and Adam sees equal amounts of Astra's life appearing in 'slow motion picture' as 'fast motion picture'. Both parts of the video signal are consistent with time dilation. If it were not present, then Adam would still receive some signals at low frequency and some at high (owing to the changing distance of travel from Astra), but in all of them Astra's movements would appear faster than Relativity predicts (and similarly for Astra's observations of Adam).

8.5 Voyaging in spacetime: inertial motion and proper time

The crucial acceleration experienced by Astra is the one at the outer part of her journey, because this is the one that allowed her to be present at two events on Earth with less proper time (that is, her own time) elapsed than someone moving uniformly between the events. The situation is similar to one with which we are very familiar concerning spatial distances: see Figure 8.8. If you want to drive between two towns A and B, then you can take a direct route, or a long route following two sides of a triangle. Although turning the corner of this triangle is only a small part of the total journey, by introducing a change of direction it requires the rest of the road to be a lot longer.

In the case of spacetime the triangular 'journey' is *shorter*, not longer, than the straight one, when we measure it by the total proper time elapsed. This is because we are measuring time not distance, and because of the mathematics of spacetime intervals.

Let us look into this. Recall that the proper time elapsed between two events at a given physical object (such as a rocket or the Earth), is the time interval in the rest frame of that object. In the case of the twin paradox we have a loop in spacetime made of two parts: the worldline of the Earth and the worldline of the rocket. Note that both parts, being worldlines of physical objects, consist completely of time-like spacetime intervals (recall the definitions in section 6.2.2). It would be interesting to ask some general questions about such paths in spacetime. What is a good way to measure the 'path length' of a path in spacetime? If we want to go from event A to event B in its future light-cone, what path or paths will be 'shortest', and what 'longest'?

Keep it clear in your mind that we are here talking about a 'path' in spacetime, not in space. Such a path does involve spatial displacements, but to measure its 'path length' we must take into account the time as well. The trouble with simply adding up all the distances or time intervals involved is that these can depend on the frame of reference that was chosen. However, a clever choice is to select, for every tiny portion of the path, the frame of reference in which the particle is at rest for that moment. This makes all the distances zero, and all the times proper times. So a natural measure of 'path length' for a worldline is the total proper time along the worldline.

We will now explore a crucial property of this measure: *proper time along a worldline is an intrinsic property of the worldline itself*. It is not like the length of an object or time separations in general, which might be subject to differing amounts of Lorentz contraction or time dilation, depending on the reference frame. Rather, we have built in to the very definition of proper time the idea of frame-independence, because no matter what reference frame may be adopted to examine some motions or events, the proper time of a particle says 'I do not care what frame you are thinking about at the moment, my proper time, between two events on me, is how much the time interval would be if you were to pick the frame in which I am at rest.' So for example, I could track the motion of a satellite by plotting its trajectory through the rest frame of the Earth, or by plotting its trajectory through the rest frame of a meteorite moving fast through Earth's atmosphere. I would derive sharply differing trajectories. However, the proper time between two events on the satellite, such as a camera shutter opening and closing, cares nothing about that, because it is a measure of time in the satellite's own rest frame: it comes with 'reference frame pre-specified'.

A general worldline can be specified by using coordinates in some chosen inertial reference frame S, and giving the position as a function of time in that reference frame. For example, this could be $x = vt$ for a particle moving at fixed velocity, $x = at^2/2$ for a uniformly accelerating particle, $x = a\sin(\omega t)$ for oscillatory motion, and so on. For a section of worldline where the particle's velocity is constant, we already know how to calculate the proper time: the relation $t = \gamma\tau$ must be applied, where t is the time elapsed in the reference frame S, and τ is the proper time. A general worldline can be broken up into many very small segments, so that in each segment the particle's velocity is approximately constant. For the segment extending from t to $t + \tilde{t}$ for some small time interval $\tilde{t}$, the proper time just for that segment is

$$\tilde{\tau} = \frac{\tilde{t}}{\gamma} = \sqrt{1 - v^2/c^2}\,\tilde{t}$$

Therefore,

$$\begin{aligned} \tilde{\tau}^2 &= \tilde{t}^2 - (v\tilde{t})^2/c^2 \\ &= \tilde{t}^2 - \tilde{x}^2/c^2 \end{aligned} \tag{8.2}$$

where $\tilde{x} = v\tilde{t}$ is the distance moved by the object in the given reference frame S, during the short time $\tilde{t}$ in that reference frame.[1]

By adding up all the little $\tilde{\tau}$ pieces we can now find the total proper time along any worldline that has been specified in some reference frame.

One interesting thing to note is that when $\tilde{x} = c\tilde{t}$ we obtain $\tilde{\tau} = 0$. This happens when the particle in question is moving at the speed of light. So any light-speed portions of a worldline contribute nothing at all to the proper time along the worldline. In particular, light-in-vacuum worldlines, which are all at light

[1] If you are familiar with the Δ notation for small quantities then you can, if you prefer, write $(\Delta\tau)^2 = (\Delta t)^2 - (\Delta x)^2/c^2$.

speed, have zero proper time along their entire length. In this sense, photons live in a sort of 'continuous present' or 'suspended animation', visiting all the places they reach without having any time elapsed at all.

So, the 'shortest' paths in spacetime, as measured by proper time, are photon paths. These zigzag through spacetime, with no elapsed proper time at all.

Now let us consider the 'longest' paths. In the twin paradox, the longer proper time is the one accumulated by Adam, the stay-at-home twin. This gives us a clue, that perhaps the longest proper time paths are the straight ones. This is indeed true, and we can prove it quite easily by appealing to time dilation. We are looking at time-like spacetime paths (worldlines) between two given events A and B. Our proposed candidate for the longest proper time is the worldline straight from A to B. This is the worldline for constant-velocity motion. Let us adopt the reference frame moving at this velocity, and consider other worldlines from the perspective of this reference frame. Some other candidate worldline must set out from A and must finish at B, which means it has only got a time interval $t_B - t_A$ to 'play with' in its efforts to accumulate lots of proper time. Unless it is the worldline with which we started, this other candidate must depart from the straight worldline at some point. As soon as it does so, it uses up some part $\tilde{t}$ of the total time available while only building up $\tilde{\tau} < \tilde{t}$ proper time (because of time dilation). Can it compensate for this by getting its proper time to increase by more than the reference frame time at some other moment? No: because time dilation only ever works in one sense: $\tilde{\tau}$ is never bigger than $\tilde{t}$. It follows that the original straight worldline is the one having the most proper time.

We now have a connection between proper time and the shape of worldlines and inertial motion:

Of all worldlines between given events A and B, the one with the most proper time is the straight one, and such a worldline describes inertial motion.

This connection proves to be crucial in General Relativity. Keep in mind that inertial motion is the motion obtained when no forces act. One of the central ideas of General Relativity is

> **Axiom of greatest proper time:** When no forces except gravity act on a particle between given events *A* and *B*, then its worldline is the one with the most proper time elapsed.

Here you see a little glimpse of the way General Relativity works: it says that the way to figure out what gravity does is to consider how it influences the rate at which time progresses. If there is no gravity, then time progresses uniformly, independent of position, and in this case we know which worldline is the one with the most proper time: it is the constant velocity one. That is the conclusion of the study we completed just above. But this means we can *derive* Newton's first law from the axiom of greatest proper time!

Next, consider a region of space where there is a gravitational field present. Then the axiom can be expressed in the equivalent form:

> *Of all worldlines between given events A and B, the one with the most proper time gives the trajectory of a body in free-fall.*

In the description of gravity offered by General Relativity, it is found that clocks run faster when they are positioned higher up in the gravity field; that is, further from whatever object to which they are gravitationally attracted. This has been experimentally confirmed. Therefore, to find the trajectory with the most proper time we should consider trajectories which move away from local gravitating objects. However, the constraint is that the trajectory must start and finish at given events—at a given place and time. This means that an attempt to get very high in the field, and thus evolve quickly, requires some high-velocity motion in order to get away and still come back in time to be present at event B. But high-velocity motion involves time-dilation, and thus *less* proper time. A particle which 'tries to' accumulate the most proper time therefore has to make a compromise between two effects:

time slow-down associated with velocity, and time speed-up associated with getting away from gravitating objects. The resulting trajectory—the one with the most proper time—represents a balance of these two, and it is the one we observe every time we toss an object into the air! The well-known parabolic trajectory of objects falling in Earth's gravitational field is precisely the trajectory predicted by the axiom of greatest proper time!

You can see now that the twin paradox, by making us think about proper time elapsed along non-straight worldlines, has opened the way to a profound insight about motion in general, and especially free-fall motion. Once you get the idea you can apply it to all types of free-fall motion. For example, consider the motion of the Earth around the Sun. For 'event A', choose some moment when the Earth is at some given location in space. Don't worry about the velocity or direction of the motion for a moment, just pick an event. Now choose another event known to be on the worldline of Earth: for example, the location at the same radius on the opposite side of the Sun, at a time six months later. Next, try to imagine all worldlines that could in principle join these two events. You could imagine, for example, firing rockets across the solar system, and plotting their worldlines on a spacetime diagram. Next, tot up the proper time along all these worldlines. Those that approach close to the Sun will not have much proper time, because the Sun's gravity slowed their clock. Those that go anywhere too fast will also not age much. A full calculation shows that the worldline with the most proper time is the elliptical (near-circular) orbit that the Earth does in fact follow!

So here is a handy tip. If you have some difficult examinations to sit at some time in the future, and you would like to give yourself the most time to prepare, what should your strategy be? Answer: build a moving trolley to transport you at constant velocity from where you are now to the place of your examination, arriving just when the examination is due to begin. Or if you are in a gravity field, you can do better still: have yourself fired upwards on a high trajectory, and arrange a suitable crash-mat, so that after a single continuous piece of free-fall motion you will land in the examination room at the time of the exam!

8.6 The glider and the hole

Our last paradox takes a look at motion in two spatial dimensions.

A flat table has a hole in it, and a stick is falling at constant velocity towards the hole, gliding along a diagonal path, as shown in Figure 8.10. In the rest frame of the table, the stick and the table surface are parallel at all times.

Suppose the hole is of diameter $D_0 = 5$ cm in its rest frame. The rest length of the stick $L_0 = 10$ cm, but its speed relative to the hole is sufficient that, owing to Lorentz contraction, it is much shortened and will just pass through the hole.

When I was young, many school desks did in fact have a hole in them: it served as an ink well or a place to hold an ink-pot. I guess that now such a hole would be used to pass computer

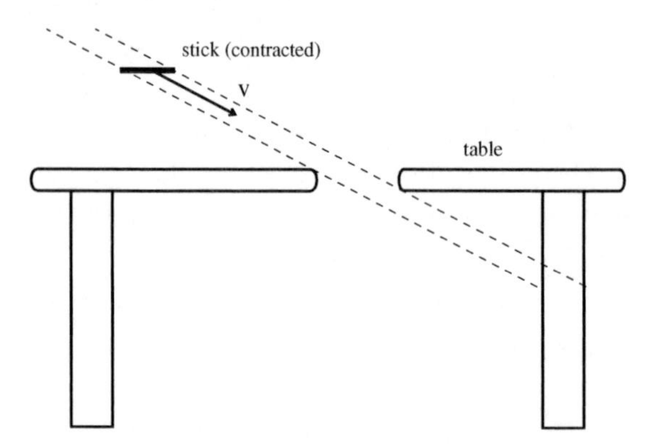

Figure 8.10: A stick falls at constant velocity towards a table. The situation in the rest frame of the table is shown. (This is not a spacetime diagram. It is a snapshot of the state of affairs in space at some instant of time in the rest frame of the table). The rest length of the stick is larger than that of the hole in the table, but owing to its Lorentz contraction the stick is shorter than the hole in the reference frame adopted for this picture. Therefore, its diagonal path, as shown, carries it right through the hole. What is the situation in the rest frame of the stick?

cables. An ordinary ruler would serve as a suitable stick, and many a schoolboy has no doubt experimented with throwing a ruler into the hole. In this example we challenge the schoolboy to throw the ruler on a gliding path, like landing an aeroplane, extremely fast, with the aim that it enters the hole while remaining parallel to the table top.

Our schoolboy is perturbed, however, by the following argument. In the rest frame of the stick, the stick obviously has its rest length 10 cm. In this reference frame it is the table that is in motion, so every part of the surface of the table is contracted, including the hole. (If you are not sure about contraction of the hole, then imagine that a ring is painted around it on the table top: this ring will shrink, but it always surrounds the hole, so the hole must shrink too.) So in the rest frame of the stick, the hole's diameter is less than 5 cm. Therefore, the stick cannot possibly fit into the hole as required! Which is true? Does the stick fall through, or does it hit the table? It cannot do both.

This example is interesting because there is no opening or closing of doors involved, as was the case in the paradox of the pole and the barn. However, like many of the supposed paradoxes of Special Relativity, the answer lies once again in a spacetime analysis and the correct consideration of simultaneity.

Resolution. When events are analysed correctly, it should not matter which reference frame we choose: both should agree on whether the stick passes through the hole. The scenario was introduced as in figure 8.10, which shows the situation in the rest frame of the table. Here it is clear that the stick does pass through the hole. We should expect that the same conclusion is reached if we consider the rest frame of the stick. Therefore let's consider the events from the perspective of that rest frame.

The crucial fact is that length contraction happens only along the direction of motion, not transverse to it. In passing from the rest frame of the table to the rest frame of the stick, the table will acquire a Lorentz contraction and the stick will lose the Lorentz contraction it had, so becoming longer. However, both these changes are *only along the direction of relative motion*, and

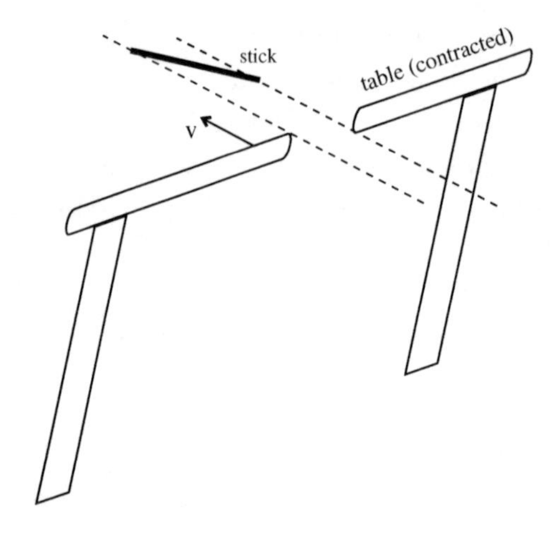

Figure 8.11: The stick and table, but now illustrated in the rest frame of the stick. The stick regains its proper length, so is elongated compared to Figure 8.10, while the table is now contracted. These changes are along the line of relative motion, resulting in a change of orientation of the stick and of the table-top, and distortion of the table. The stick still passes through the hole.

they therefore have no influence at all on whether the stick passes through the hole: see Figure 8.11. Suppose we arrange a tube that just passes through the hole, oriented so that the stick falls down it with precisely the same motion and orientation as before. Then the Lorentz contraction only affects the spatial extent of any object along the direction of this tube, not perpendicular to it.

The Lorentz contraction in this example results in a change in the angle between the stick and table-top, in such a way that the stick still just passes through, even though it is longer and the hole shorter. A good way to make this change of angle obvious is to consider a rectangular block sitting on a fast-moving train. Suppose a line is painted diagonally on the side of the block, from the top right to bottom left corner: see Figure 8.12. If the block is contracted by Lorentz contraction, then its length is reduced but its height is unchanged. Therefore the painted diagonal line between opposite corners must get steeper. The same must happen to a steel strut attached diagonally across the block—and therefore the angle between such a strut and the train track will

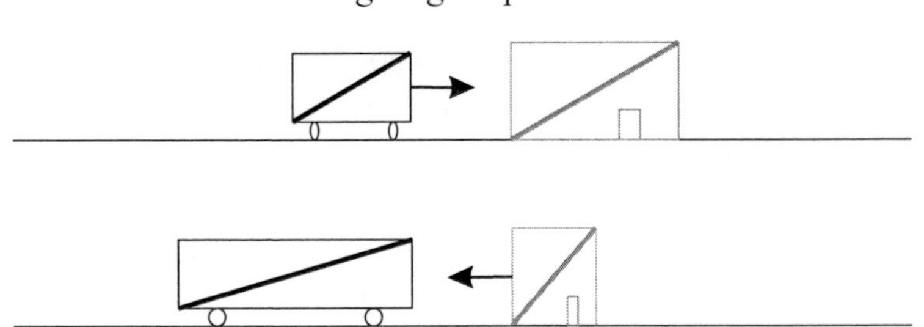

Figure 8.12: The angle changes of diagonal lines when objects undergo Lorentz contraction.

Figure 8.13: Spacetime diagram for the glider and hole paradox. The motion is in two spatial dimensions, resulting in a three-dimensional spacetime diagram. The diagram shows the motion of the stick (the dark grey worldsheet) and a cross-section through the table top (light grey worldsheets). The two dotted events are simultaneous in the rest frame of the table; at that moment the stick just fills the hole. This implies that the stick is parallel to the table-top in the rest frame of the table. In most other reference frames the dotted events are not simultaneous, so the stick passes through the hole at some angle to it.

depend on the reference frame. This will happen whether or not the rectangular block is present.

A spacetime diagram for the glider and hole scenario is shown in Figure 8.13. The spacetime diagram is three-dimensional, because we need two spatial dimensions to characterize the motion. It is intended to give you a general impression of what happens. The gliding stick is represented by a worldsheet, and another worldsheet shows the particles making up one cross-section of the table-top. This sheet has a hole in it. The motion of the stick means that the stick's worldsheet is angled relative to the hole, and it just fits through. Therefore, the stick passes through the hole, no matter what reference frame you choose. This diagram is useful because it makes it clear that there are not two alternative scenarios, depending on the reference frame: there is just one set of events in spacetime. The sequence of events will depend on how we 'slice up spacetime' by adopting one reference frame or another, but this is no more mysterious than the fact that cross-sections of a cake will look different if you slice it in different ways.

8.7 The birthday party

Let us finish this chapter by solving an example of the kind of problem that can come up in Special Relativity. Suppose Astra leaves Earth on her twentieth birthday, accelerating quickly up to three fifths of the speed of light (relative to Earth) and then coasting at this speed. On her next birthday she sends a message back to Adam, inviting him to join her for her twenty-fifth birthday party. Does Adam receive the message in time for him to make the rendezvous? If so, how fast will he have to travel to catch up with her and not miss the party?

Here is how to answer these questions. At $v = (3/5)c$ we have $\gamma = 5/4 = 1.25$. We assume that Astra celebrates each of her birthdays when she has aged by a year, so owing to time dilation Adam finds that 1.25 years pass on Earth between Astra's birthdays. Therefore, when she sends the message her distance from Earth is $1.25v = 0.75$ light-years, in the Earth rest frame. It follows that in Adam's rest frame the message takes 0.75 years to travel, and Adam receives it two Earth-years after Astra set out. These facts lead to the lower left part of the spacetime diagram shown in Figure 8.14 (or you could refer to Figure 8.9b).

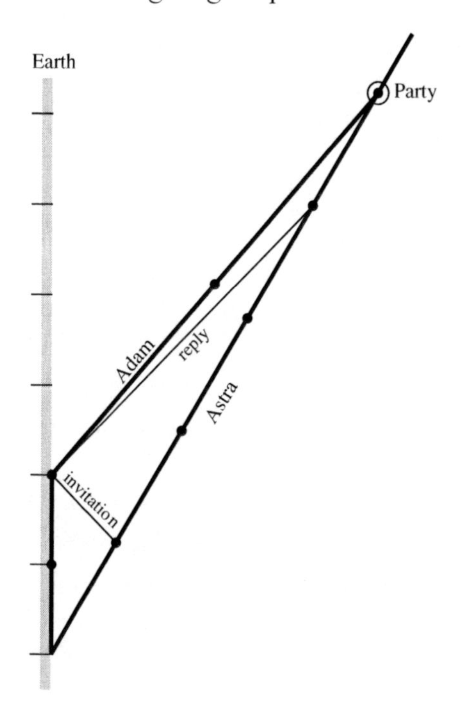

Figure 8.14: Spacetime diagram for the birthday party arrangements.

Astra meanwhile finds Adam's life to be time-dilated, so she calculates that he receives the message not two years after she left him, but $\gamma \times 2 = 2.5$ years after. He is then $2.5v = 1.5$ light-years away, so his reply takes 1.5 years to reach her, and she receives it on her twenty-fourth birthday (see Figure 8.14 again, and Figure 8.9a also illustrates this). Clearly Adam can get to the party, because he doesn't have to travel faster than light.

Before he leaves, Adam knows that the occasion of the party is to be at Astra's time five years after she set out, and therefore at Earth time $5\gamma = 6.25$ years, when Astra will be $6.25v = 3.75$ light-years away. He has $6.25 - 2 = 4.25$ years (in Earth's rest frame) to get there, so he needs to travel at $3.75/4.25 = 15/17 \simeq 0.882$ times the speed of light, relative to Earth.

As a check, let us use equation (7.11) (addition of velocities) to find out Adam's speed relative to Astra. Note that we need to have the sign correct: Earth's velocity relative to Astra is $-v$. Therefore, the calculation is

$$\frac{15/17 - 3/5}{1 - (15/17) \times (3/5)} c = \frac{3}{5} c$$

for the speed of Adam relative to Astra. (Note that previously he was moving away from her at this speed, but now he is moving towards her). In Astra's rest frame Adam must cover the same distance of 1.5 light-years that his reply message has to travel, so it will take him $1.5/0.6 = 2.5$ years of Astra's rest frame time. This is just the right amount for him to arrive at the party, so everything checks out.

When they meet, Adam will be younger than Astra, of course, because now we have the situation of the twin paradox again, but this time it is Astra who took the inertial path, and Adam who has the kink in his worldline.

Puzzle How old is Adam at the party?

9

Faster than light

In this chapter we will revisit the Light Speed Postulate, in order to clarify the relationship between version A and version B, and to discuss various scenarios where 'faster-than-light' propagation is concerned.

The Light Speed Postulate does not claim that no one can even conceive of a speed faster than light. It claims merely that no causal influence can propagate faster than light (to be precise, than the speed of light in vacuum). For example, when you look up at the stars, at one moment you may be looking towards the 'pointers' in the constellation Ursa Major (Merak at 79 light-years distance from Earth, and Dubhe at 124 light-years), the next you may sweep your gaze up to the pole star (Polaris at distance 430 light-years). You have thus moved the point towards which you are looking by a distance of about 300 light years in around half a second: a speed approximately ten billion times the speed of light! Special Relativity, via the Light Speed Postulate, places no restrictions on 'speeds' such as this one. No physical object or influence was associated with the movement of the point you were looking at.

9.1 Faster than light list

We will now provide a list of examples of things which may at first appear to break the Light Speed Postulate, though on reflection it is clear that they do not (see Figure 9.1). In each case a speed

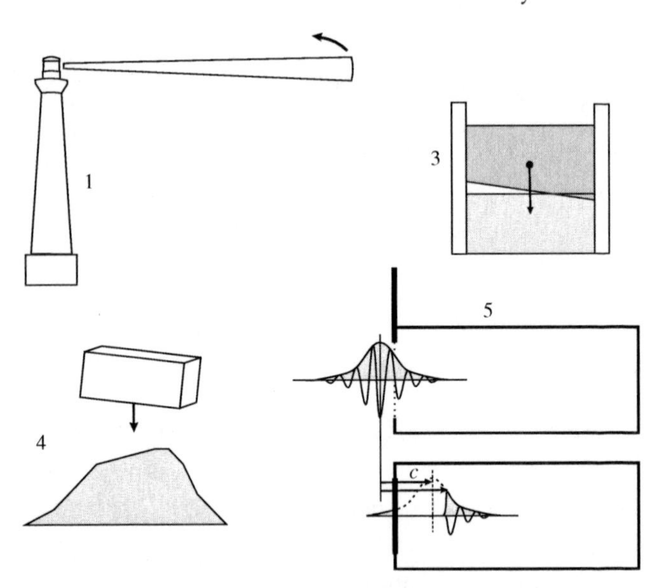

Figure 9.1: Some examples of physical phenomena involving a speed faster than light.

faster than light can be defined, but no signal is propagating at this speed. As a test of this, try to imagine using each phenomenon in the list to send a message from one place to another. This will be discussed after the list has been presented. For the examples concerning wave motion, recall that the speed of the wavefronts of a wave of frequency f and wavelength λ is always $f\lambda$.

1. *Lighthouse beam.* A beam of light from a lighthouse falls on a wall at distance r from the lighthouse. The lighthouse lamp rotates, completing a revolution once every period T. If the wall were to encircle the lighthouse, then its circumference would be $2\pi r$. The illuminated spot must move around this circumference in time T. It follows that the illuminated spot moves across the wall at a speed $2\pi r/T$. There is no restriction on r or T: this speed is unrestricted.
2. *Shadows and projection.* Similar to the lighthouse is the case of a moving shadow cast by a moving object on a screen far away. If the light source is small and emitting in all directions, then the shadow of an object moving around it or near it behaves just like the lighthouse beam in the previous example. A related idea is that

of projection, as in a movie theatre. If an image of Buzz Lightyear is projected onto a screen, the pattern of light and shade can be made to move across the screen at any speed. The speed attained by the projected image of Buzz can go to infinity (and beyond!).

3. *Guillotine.* A guillotine has two straight blades. They maintain a fixed angle θ to one another while one blade moves down at speed v. When the ends of the blades are separated in the vertical direction by distance y, the point at which the two blade edges cross is at a fixed height and at horizontal location x given to good approximation by $x = y/\theta$ when the angle θ is small. The rate of change of y is given by the speed of movement of the upper blade v, so the rate of change of x is v/θ. In other words, the crossing point of the blades moves in a horizontal direction at speed $w = v/\theta$. v is restricted: it cannot exceed c, but θ can be very small. For small enough θ, the horizontal speed w can exceed the speed of light, and as θ tends to zero, w tends to infinity.
4. *Brick and sand-hill.* A child makes a small mound of sand on the beach. The mound slopes gently upwards to a peak at one end. The child drops a brick on the sand, the brick being angled in the opposite sense to the mound, so that when it is removed the mound has been depressed into a shape sloping to a peak at the other end. The horizontal length of the brick is d, and its speed relative to the mound is v. Its angle is such that the final slope in the sand has a vertical change of height h over the horizontal distance d. The time taken for the brick to depress the sand is approximately h/v. The peak in the sand moved through a distance d in this time. The speed of movement of the peak is therefore dv/h. Since h can be small, this speed is unrestricted.
5. *Shuttered light pulse.* A light-pulse a few metres long begins to arrive at a large telescope. The peak of the pulse moves as $x = ct$ where x is the distance from a shutter at the front of the telescope. At time $t = -T$, where T is less than half the pulse width, the shutter is closed. Some light has entered the telescope, and continues to propagate. Its peak intensity lies near $x = 0$ just after the shutter closure. Therefore, the closing of the shutter causes the peak of the pulse (the maximum intensity point) to move from $x = -cT$ to $x = 0$ in a very short time. This movement of the peak is much

faster than c. At later times the peak is at $x = c(t + T)$. Between times $-2T$ and $2T$, for example, the peak moves overall from $-2cT$ to $3cT$, representing a distance of $5cT$. The average speed of movement of the peak during this time interval is therefore $(5/4)c$.

6. *Fast attenuated pulse.* In some circumstances light pulses can propagate in a medium such that the peak of the pulse moves faster than c, while typically the whole pulse is attenuated.
7. *Primed Mexican wave.* If a crowd of people, such as at a sporting or musical event, are in the mood for some fun, then sometimes they like to do a 'Mexican wave'. This is when each person stands and waves their arms and sits down again, just after their neighbour has done the same. The result is a wave of movement running through the crowd, which travels typically much slower than the speed of light. However, if a child first ran through the crowd, priming people to get ready to wave, then he could arrange a faster-than-light Mexican wave, as follows. Suppose each person is at a distance of two light-seconds from his neighbours (which is a long way—but we imagine an impractical example just to make the point). The child runs through the crowd, telling each person he meets a number. This is the number of seconds after eight o'clock that the person has to stand and do his wave. The child tells the first person '1 second after eight', the next person '2 seconds after eight', the next person '3 seconds after eight', and so on. When eight o'clock arrives, the Mexican wave starts, and it travels across the crowd at twice the speed of light, because each person waves one second after his neighbour, while light takes two seconds to pass between them. With different primed instructions, any speed could be arranged.
8. *Fast amplified pulse.* In a suitably prepared medium, a light pulse can move faster than c while maintaining its strength or even growing by amplification.
9. *Electromagnetic waves in a waveguide.* A waveguide is a tube made of some material that strongly reflects electromagnetic radiation (such as a gold-coated tube for microwaves, or an optical fibre for light-waves). Electromagnetic waves can propagate down such a tube. For given frequency f, the wavelength λ in the guide is related to the wavelength λ_0 in free space by the 'waveguide equation'

$$\frac{1}{\lambda^2} = \frac{1}{\lambda_0^2} - \frac{1}{\lambda_c^2}$$

where λ_c is a cut-off wavelength that depends on the shape and physical dimensions of the waveguide; the result only applies when λ_0 is smaller than this. Therefore, the speed at which the wavefronts move down the waveguide is $f\lambda = c/\sqrt{1 - \lambda_0^2/\lambda_c^2}$. This is always larger than c, and tends to infinity as λ_0 tends to the cut-off wavelength.

10. *de Broglie waves.* In quantum mechanics, particles are associated with waves of wavelength $\lambda = h/p$ and frequency $f = E/h$ where $h = 6.626 \times 10^{-34}$ Js is Plank's constant, p is the momentum of the particle, and E is its energy. Waves of wavelength λ and frequency f must move at speed $f\lambda$, so the speed of these waves is $w = E/p$. In the presence of potential energy, E can be non-zero while p is very small. Hence the speed of the wavefronts is unrestricted.
11. *Closing speed.* Two light-pulses approach a given location A—one from the left, and one from the right. Each travels at the speed of light, c. An observer at A must conclude that the distance between the light pulses is diminishing at the rate $2c$. Similarly, particle beams in an accelerator may approach an observer at the target location from opposite directions, both at close to the speed of light. Such an observer finds the distance between the particles diminishes at close to $2c$.
12. *Classical correlations.* Two sets of cups are located on the planets Earth and Mars. Under each cup is a fair coin. Each five seconds, a cup on Earth is lifted, and the corresponding cup on Mars is lifted one second later (in the rest frame of the solar system). It is found that the coins on Mars have precisely the same sequence of heads and tails as the coins on Earth. (The distance from Earth to Mars ranges from 3 to 21 light-minutes).
13. *Quantum path calculations.* In the calculation of the behaviour of particles in quantum mechanics, a summation is carried out, involving all possible trajectories of the particles, in order to find the probability that a particle will pass from one place to another between given times. It is not necessary to leave out of the sum trajectories involving faster-than-light propagation.

14. *Gravitation and cosmology.* Gravity has the remarkable property that in every small region of space, the speed of light (measured in terms of local clocks and rods of some given construction) is the same, and yet there is an effect of gravity on the propagation of light. The effect can be understood in terms of the geometry of curved surfaces, applied to spacetime itself; this is the essential idea of the theory of General Relativity. In this situation the definition of *speed*, whether of light or of anything else, is quite subtle. In the presence of gravity, an entity such as a particle or a signal still cannot travel faster than the speed of light in its immediate vicinity, but when a coordinate system appropriate to some region A is extended to another region B, there can be an apparent change in the speed of light, between one region and another. In this situation it is easy to make ill-defined statements about speed (or to find them on the Internet!).

Discussion

Some of the examples in this list are straightforward, some require further thought, and some are preliminary to a wider discussion.

1, 2. The lighthouse beam does not transmit information from one patch of wall to another, but from the lamp to the wall. Nothing anyone might do at the location A of the illuminated patch of wall at any instant can have the slightest effect on what will happen at the piece of wall B that the patch happens to illuminate next. The light from the lighthouse is already in transit and will arrive at B quite undisturbed by events at A. Similar statements apply to a projected image.

3. The guillotine example is similar. No propagation of information is represented by the movement of the intersection point of the blades: it is merely a mathematically defined point. If something were to impede or change the movement, such as bump on the blade, then causal information would propagate. However, the deformation resulting from the blade's hitting such a bump would propagate down the blade much more slowly, at about the speed of sound.

4, 5, 6. The child on the beach is similar to the guillotine, and also serves as a prelude to the examples of chopped or attenuated light pulses. The shutter on the telescope appears to make the light pulse 'leap forward,' but of course the part of the pulse that forms the new maximum was already inside the telescope, and it did not leap anywhere. The attenuated light pulse examples succeeded in catching out some scientists who should have known better, and who claimed to have demonstrated faster-than-light propagation of information using it. For example, one could play a Mozart symphony, have the signal transmitted by fluctuations of a laser beam, allow the laser beam to propagate through a medium where pulses propagate faster than c while being attenuated, then detect the light on a photodiode, amplify the fluctuating current, and use it to play back the music. The beautiful music arrives at the output faster than a light-speed signal could have carried it. Or does it? This appears very much like a violation of the Light Speed Postulate, but before leaping to that conclusion one needs to check whether the medium is functioning like the example of the shutter on the telescope. The light detector produces a fluctuating output in proportion to the peaks and troughs of the light intensity it receives, and each such peak can arrive before the peak of an unattenuated beam would have done. However, with the hint provided by the shutter example, we can soon convince ourselves that the minimal information required to reproduce the music was already in the weak leading edge of each peak, which did not propagate faster than light.

7, 8. The special Mexican wave does not transmit information. The information was transmitted by the child who set up the wave and who travelled slower than c. In the case of a fast pulse of light in a specially prepared medium, something similar takes place.

9, 10. The examples of wavefront speeds for waves (the technical term for this is 'phase velocity') are all dealt with by the fact that a wave having a single precise frequency and wavelength is a mathematical abstraction. It refers to a wave which is perfectly predictable: its physical extent is infinite, and it continues to oscillate for ever. Therefore, the arrival of each wavefront does not constitute arrival of information. The wavefronts are 'expected'. Real waves consist of a finite duration of oscillatory motion, and can be described as a set of ideal waves with a range of frequencies and wavelengths. Any *change in the wave motion* instigated at some position does *not* propagate down the wave at the speed of the wavefronts, but at another speed called the 'group velocity' (the group velocity is equal to $df/d\lambda$, the rate of change of frequency with wavelength). The group velocity in these examples is less than c.

11. The closing speed example was discussed in Section 7.3. For calculations in any given reference frame, the closing speed is a perfectly respectable and useful concept. One simply needs to understand that it is not the speed of propagation of any physical entity or information, and it should not be confused with 'relative speed', which is the speed of one body in the rest frame of another.

12. The example headed 'classical correlations' is a prelude to the following brief introduction to a subtle quantum effect.

The example as given involved purely classical physics, with coins and cups. The description is intended to invite a sufficiently muddled person to imagine that some sort of influence must propagate from Earth to Mars faster than light, to achieve the observed 'amazing' agreement between the two sets of coins. Of course, nothing of the sort is needed. The coins could have been set up some time ago. Nothing happens to them when the cups are lifted. The investigators simply become aware of which way up their coins are. The investigator on Earth cannot know that the coin on Mars is the same way up as his own until sufficient time (3 to 21 minutes) for the information to reach him has elapsed.

A remarkable quantum effect is related to this, but is more subtle. It goes by the various names of 'Bell-EPR experiment', 'entanglement', and 'quantum correlation'. Quantum mechanics permits that a set of pairs of physical entities such as atoms can be prepared in a special state, such that a sequence of different sorts of observations of them exhibits correlations, like the correlations 'head–head' and 'tail–tail' of the classical coins. However, for the quantum 'entangled' state, it is not possible to explain the degree of observed correlations, for all possible measurements, simply by imagining the atoms were prepared with pre-existing states and were merely observed without affecting them. Rather, such entangled pairs exhibit, to a partial extent, a shared identity, such that operations on one atom cannot be distinguished from operations on both. However, the resulting correlations are always of the sort illustrated in the coin example. The observations at any one atom are random, and it is only when results from the two locations are brought together and compared that the correlation can have any influence on further physical phenomena. This bringing together requires communication, and is limited by the light-speed limit. Therefore no faster-than-light signalling can be achieved by making use of these quantum correlations. However, they can be used for other things: they are at the heart of an area of physics called *quantum information theory*, and they allow a powerful form of information processing, called *quantum computing*, to be implemented.

13. The penultimate example—quantum path calculations—is somewhat like the issue of wavefront velocity and group velocity. The calculations in quantum physics involve mathematical methods that can be interpreted as including a faster-than-light contribution, but when the predictions for physically observable phenomena are arrived at, no violation of the Light Speed Postulate is found.

14. Finally, the subject of gravitation and cosmology involves warped spacetime where the whole notion of speed has to carefully reconsidered. It would take another book to explain what is meant when people talk somewhat loosely of 'faster-than-light' propagation in this context. However, let me offer you the following simple observation. Suppose an ant is walking along an elastic band, while someone continuously stretches the band. For example, suppose that each second the ant can travel 1 centimetre along the surface of the rubber, while the length of the band is increased by 10 centimetres. The ant sets out from a drawing pin attached to one end of the band, towards some sugar lying on the other end. With the far end of the band moving away so quickly, you might think that the ant will never reach the sugar. However, it does.

To prove this, it helps to simplify the problem a little. We suppose that the stretching of the band and the crawling by the ant take place alternately. First the band length increases by 10 cm, then the ant crawls 1 cm, then the band is stretched again by 10 cm, then the ant crawls another centimetre, and so on. Think about the fraction of the length of the elastic band that the ant has covered. Let us call this fraction p. At the outset, $p = 0$. After the first stretch it is still zero, but then the ants crawls a little and covers some small fraction p of the length of the band. Suppose for example that the band length is now 20 cm, so $p = 1/20 = 0.05$. The important point is that the next stretch does not diminish the fraction: the ant is carried along on the surface of the rubber, and is now one twentieth of the way along the new band length of 30 cm. It has further to go than it had (it had 19 cm to go, and now it faces 28.5 cm), but like the tortoise it does not lose heart, and scurries another centimetre, to gain a total distance from the drawing pin of 2.5 cm, so now $p = 2.5/30 = 0.08333$. The band then extends again, leaving p unchanged, then the ant scurries again, increasing p from $2.5/30$ to $2.5/30 + 1/40 = 0.108333$. You see the pattern: at the end of each crawl, the ant always manages to increase the proportion of the whole length that it has covered. The question is, does p ever increase all the way to 1, or does it merely approach closer and closer to some smaller number such as 0.2 or 0.5? You can check this by examining the sequence of values of p. Each value is larger than the previous one by $1/L_n$, where L_n is the length of the elastic band at the nth step, in centimetres. The following table summarizes the results.

After 33, 615 stretches, the length of the elastic band is 3.3616 kilometres, the ant has been travelling for over 9 hours, and it has 0.404677 cm left to travel. The next stretch increases this to 0.404689 cm, which the ant easily covers, and it then eats the sugar.

In this little parable the ant can be compared to light propagating across the universe. Just as the legs of the ant can only pull it along at a fixed maximum speed relative to the rubber, so also light can only travel at a fixed maximum speed relative to

n	L_n	p_n
0	10	0
1	20	$\frac{1}{20} = 0.05$
2	30	$\frac{1}{20} + \frac{1}{30} = 0.08333$
3	40	$\frac{1}{20} + \frac{1}{30} + \frac{1}{40} = 0.10833$
...		
10	110	0.20199
100	1010	0.4197
1000	10010	0.6486
10000	100010	0.8787
33615	336160	0.9999987

objects in its vicinity. The atoms in the rubber may be compared to galaxies in the universe. The stretching of the band illustrates the cosmological expansion of the universe. One might say that the sugar moves away at 'faster than the speed of ant', yet the ant still catches up with it. In the reference frame of the drawing pin, the ant itself travels at 'faster than the speed of ant'. You can decide for yourself whether you think this is a helpful way to express it.

9.1.1 TACHYONS

In theoretical physics it is a useful practice to introduce, for purposes of discussion, physical behaviour not known to be possible, or even believed to be impossible. The idea is to explore theoretically what would follow from such behaviour, and to ask whether it might be possible after all.

An example of this is the notion of 'tachyons'. This is the name given to particles that propagate faster than light. No such particles have ever been observed, but attempts have been made to explore whether they might be possible, and if so, what they would be like. For example, one can put forward arguments to suggest that for such a particle c would be the minimum instead of the maximum attainable speed.

However, these attempts to find a physically sensible role for tachyons have so far failed, and it is believed that tachyons are simply not possible.

9.1.2 SLOWER THAN LIGHT

One can find examples of physical phenomena where a misunderstanding might lead one to think the Light Speed Postulate was violated, but in fact the relevant speeds are all less than the maximum allowed, so no violation occurs. It is worth taking a look at some of these for the sake of clarity.

The first two concern the propagation of light inside a material medium such as glass or water or a gas. If n is the refractive index of some medium, then the speed of propagating electromagnetic waves in the medium is c/n. Therefore when $n > 1$ (the usual case) these wavefronts move at a speed less than c. In this case the Light Speed Postulate does not prevent particles from moving faster than c/n as long as their speed remains below c. Such a particle is said to be moving 'faster than light in the medium', but strictly speaking the wave motion in this case is not purely light (a motion of electric and magnetic fields alone), because it also involves the movement of material particles such as atomic electrons in the medium.

1. *Cerenkov radiation.* Water has a refractive index of approximately 1.5 so the 'speed of light in water' is substantially less than c. Processes such as radioactive decay can produce particles such as electrons travelling through water at a speed close to c, and therefore faster than c/n. Their passage disturbs the electric and magnetic fields in the water, with the result that light is emitted. This light is called 'Cerenkov radiation'. Its production has some similarities with the 'sonic boom' experienced when aircraft travel faster than sound. Looking for Cerenkov radiation proves to be a useful way to detect fast-moving particles in particle physics experiments, and it also allows their speed and direction to be determined.

2. *'Slow light'*. There exist some ingenious experimental methods that allow the refractive index of a gas to become very large over a small range of frequencies. In this case the 'speed of light' in the gas (c/n) can be just a few metres per second, or can even tend to zero. This experimental achievement was celebrated with the name 'slow light', but in the circumstance of such a high value of n, the propagating waves in fact consist almost entirely of the motion of atoms or electrons. The name 'slow light' is therefore partially misleading, but it is useful because it captures the idea that the electrical motion in the medium can be converted back into light-waves at the edge of the medium.
3. *Visual appearance*. In some circumstances there can be a visual appearance of faster-than-light motion when one does not take into account the travel times of the light-waves. For example, astronomical observations might detect a ring-shaped disturbance moving outwards from some stellar explosion, heating the surrounding layers of dust, and causing them to glow. The radius of the ring might be determined as 1 light-year at some moment, and then a few months later it is measured as 2 light-years. Does this mean that the ring moved a whole light-year in only a few months? No. What is happening is that the ring is moving not purely transversally but also partly towards the viewing astronomer on planet Earth. The light that set out from the first glowing dust layer has further to travel to Earth than that from the second, so instead of arriving at Earth more than a year before the second pulse, it arrives only a few months before. After taking this travel time into account, one concludes that the ring was not in fact expanding at more than the speed of light.

9.2 Light speed is information speed

There was a young lady called Bright,
Who could travel faster than light;
She went out one day,
In a relative way,
And came back the previous night.

Printed in *Time*, July 1946

The list of examples in Section 9.1 shows that there are plenty of cases in which the position of some physically identifiable phenomenon propagates faster than light. This leads us to ask what in general is speed-limited, and what is not. Does the limit apply only to light-waves and material objects such as particles? Is there some simple general rule to guide its application?

Such questions are answered by filling out the definition of the term 'signal' that was used in the statement of version A of the Light Speed Postulate.

Let us consider again what would be 'wrong' with an influence travelling faster than the maximum allowed speed for signals. The point can be made with a colourful example with a science fiction flavour. Science fiction writers like to invoke the possibility of faster-than-light travel for spaceships, because it simplifies the plot when one is considering travel around a galaxy where stars are typically separated by many light-years. So let us suppose some advanced engineers have succeeded in developing a faster-than-light delivery system: some sort of tube that can accelerate objects inside it up to $10c$ relative to the tube. Unfortunately their home planet is attacked. The best they can do is escape with their lives on board an 'old technology' vessel that can achieve a speed of $0.8c$ relative to the planet before its fuel runs out. It takes a few hours for the attackers to learn the operation of the faster-than-light system, and they then use it to fire a golden missile at $10c$ towards the fleeing vessel. It arrives there after fifteen minutes (planet time) and explodes. Only a few hardy robots survive to tell the tale.

In this scenario the events 'missile is fired from tube' and 'missile explodes at vessel' have a space-like separation. This implies that their time ordering can depend on the reference frame. In particular, in the reference frame of the fleeing vessel, the explosion happens first (see Figure 9.2).

In the reference frame in which the escaping vessel is at rest, the sequence is as follows. Just as they are celebrating the development of their new faster-than-light delivery tube, the engineers find themselves under attack. They blast off in their escape ship,

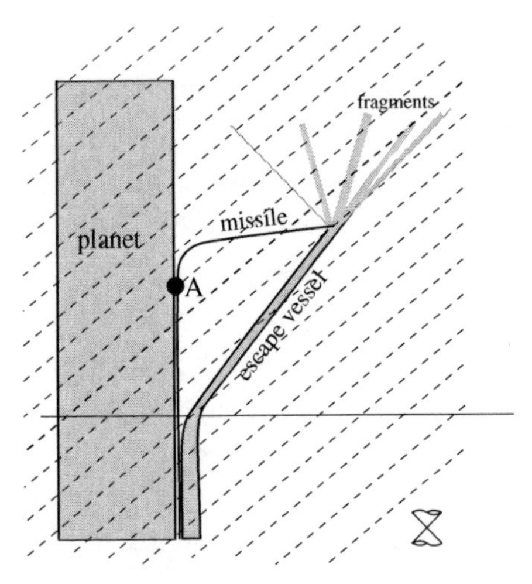

Figure 9.2: An impossible spacetime diagram. An escape vessel leaves the planet at relative speed 0.8c. Subsequently, at event A a missile launch tube is fired and a missile travels between planet and escape vessel. In the rest frame of the planet, the missile travels away from the planet at the impossible speed of 10c. It hits the escape vessel, destroying it in an explosion. The explosion results in a flash of light and other fragments. The horizontal line shows an example line of simultaneity in the planet's rest frame. The dashed lines show a set of lines of simultaneity for the rest frame of the escape vessel, at hourly intervals in that frame. It is seen that in this reference frame, the explosion of the vessel happens about two hours before A. The missile worldline represents motion of the missile *from* the vessel *to* the planet, and of a second missile launched from A. These two meet one another near the planet, and vanish.

noticing as they do that the attackers have brought some sort of shiny gold missile with them. After about two hours (vessel time), the engineer's escape vessel spontaneously explodes. The explosion produces not only debris, but also a faster-than-light missile, made of some shiny gold matter not previously present on the vessel. It propagates back to the planet. It arrives there exactly at the location of the long tube, and starts to accelerate down the tube. Another golden missile meanwhile accelerates up the tube to meet it, and just as they both achieve infinite speed relative to the reference frame of the (now fragmented) escape vessel, they meet and annihilate each other completely harmlessly.

What this illustration makes plain is that if one wants to contemplate the physical possibility of faster-than-light travel of ordinary objects, one has to reflect carefully on whether the physical scenario under discussion is plausible in all reference frames. The sudden appearance of the missile in the midst of the debris of the spontaneous explosion in the escape vessel reference frame breaks all sorts of physical principles, such as the increase of entropy and the conservation of various types of particle number. In either reference frame we have imagined that the particles of the missile could accelerate to beyond light speed, but it requires infinite momentum and infinite energy for them even to approach the 'light barrier'. In short, the whole scenario is impossible.

Faster-than-light travel in one direction is not exactly the same as time-travel backwards in time (with all its well-known paradoxes such as the grandfather paradox), but it almost amounts to that. A pair of faster-than-light signals in opposite directions can be used to create even more bizarre conclusions than those we found for the golden missile, including a form of the 'grandfather paradox' where a man travels back in time and kills his own grandfather, thus preventing his own birth, making it impossible for him to carry out the killing, and so on. What such arguments really show is that once we suppose a missile could travel faster than the maximum speed specified in the Light Speed Postulate, then we must abandon the whole theory of Special Relativity and start again. However, the argument helps us to tease out what could and could not be allowable as a faster-than-light motion. The main point is that faster-than-light motion is a form of motion in which the direction of movement can depend on reference frame. If the fast-moving entity can exert any influence on other things, then this results in a reversal in the time-ordering of cause and effect, which is logically absurd. Just one contradiction is enough to produce nonsense. The conclusion is that the maximum speed limit applies to *anything that can influence other things*.

A form of words that expresses this concept clearly is to say that the speed-limited physical property is *information*. Here we use the word 'information' to refer to anything that is detectable and that could in principle have more than one value or form. For example, think of a text message: because there are many different messages that could in principle be sent, the one that is sent constitutes information. Or consider any particle: its arrival at a point in space brings information in the form of its rest mass, charge, and other properties.

If a particle had no discernible properties, then it would not be able to influence other particles, and then it could travel as fast as it liked . . . but then neither we nor any other physical system would ever become aware of it or be affected by it, so we may as well say it does not exist.

By talking about information we also make it clear how to understand the speed-limit when a complicated series of interactions is involved. For example, a light pulse could arrive at an electronic detector, which emits a pulse of current down a cable, which triggers a particle accelerator one mile away, which emits a fast proton from the end of a five-kilometre vacuum tube. We do not need to look into all the details of a complicated sequence of events like this in order to know whether the final event of emission of the proton has to have a time-like separation from the initial trigger event of arrival of the light pulse. All we need to know is the answer to the question, 'if the light did not arrive, would the proton still be emitted?'; or more generally, 'would a change in the arriving light pulse result in a change in the emitted proton?' If the answer is 'yes', then the process is speed-limited: the interval has to be time-like.

9.3 Equivalence of two versions of light speed Postulate

Most introductory texts introduce the Light Speed Postulate in the form we have called version B, 'The speed of light in vacuum is independent of the motion of the source.' However, version A,

'There is a finite maximum speed for signals,' is slightly purer because it shows that Relativity is not dependent on a theory of light or electromagnetism; rather, it underpins that theory and all other physical theories.

We would now like to show that version B can be proved from version A, in the following sense. It is not possible to deduce that there exists such a thing as light, merely by reasoning. However, one can show that if some information-bearing entity propagates in vacuum at the maximum speed mentioned in version A, then that type of entity always propagates at the maximum speed, independent of the motion of other bodies. In this sense, version B follows from version A.

Suppose then that we adopt version A, that there is a maximum speed for signals. This could be discovered experimentally without reference to light. For example, machines can be (and have been) built to accelerate particles to higher and higher velocities, and it emerges in such experiments that there is a universal limiting speed which cannot be exceeded. Call this limiting speed c. It immediately follows from the Principle of Relativity that the same value will be found in all reference frames. It only remains to show that if any given entity has the speed c in one reference frame, then its worldline will be found to have that same speed in all other frames. This is enough to obtain version B.

Suppose therefore that an information-bearing pulse P has speed c in some reference frame S, and let c' be its speed in another reference frame S$'$ (note there is just one pulse under discussion, not two, but we can track its motion relative to as many reference bodies as we like). The speed c' in S$'$ cannot be greater than c according to version A of the Light Speed Postulate. Suppose, then, that $c' < c$. But this would mean that it is possible for a particle to move faster than the pulse in frame S$'$, so that it could catch up and overtake the pulse. Such an overtaking is frame-independent (it means the worldlines cross, which will be the case in all reference frames), so it implies that the particle moved faster than the pulse in frame S also, where the pulse speed is c. But that is ruled out by version A. Therefore $c' = c$.

This proves that something having the maximum speed in one reference frame has that speed in all reference frames, and therefore its speed in any given reference frame will not depend on the motion of the source that emitted it. Therefore we have proved version B of the Light Speed Postulate.

10

Introduction to momentum and energy

So far, our whole discussion of Relativity has focused on the nature of spacetime. We have considered time, position, velocity, and acceleration, but not mass or momentum or energy or force, except in the brief preview in Chapter 2. In technical language, we say we have discussed 'kinematics', but not yet 'dynamics'. In this chapter we will introduce the basic ingredients of dynamics in Special Relativity. This can be done completely accurately while keeping the mathematics simple. The main aim of the chapter is to answer the question, 'how is the famous equation $E = mc^2$ obtained, and what precisely does it mean?'

The physical content of Einstein's remarkable equation is that energy and mass are equivalent physical quantities. Although we have developed different words for them, they are not two properties, but one and the same property. They are no more different than 'water' and '$H_2 0$'. We have two words 'energy' and 'mass' partly because our forebears were unaware of their equivalence, but also because we have found it convenient to keep both words in our vocabulary, since they help us to focus on different aspects of the 'mass-energy'. The presence of the c^2 term shows that in everyday units such as joules and kilograms the amount of energy associated with a given amount of mass is very large: about 90 million billion joules per kilogram (9×10^{16} J/kg). This means that the total daily energy production of all the power stations in the world could in principle be obtained from just 14 kilograms of raw material. That has important practical consequences, but from a fundamental physics point of view it only amounts to saying that the energy associated with a given amount of mass is large compared to a human-devised unit of 'joule per kilogram'. We could if we wished adopt a different set of units, and then the numerical value of the 'energy per unit mass' would come out as some other number. For example, if we let the unit of distance be the light-second (the distance light travels in vacuum in a second), abbreviated to ls, then the speed of light is $c = 1$ ls/s, and the unit of energy is kg ls^2/s^2. In these units, the amount of energy per unit mass is 1 energy-unit per kg.

A more complete discussion of mass-energy is only possible once we have derived the formula and learned what its implications are for physical phenomena. That is the only way to grasp its meaning properly. The equivalence of energy and mass is, I feel, the 'jewel in the crown' of Special Relativity. It is the insight with which Einstein was himself most delighted, and of which he was understandably most proud. I know you would not expect to get to grips with a jewel of this value without a certain amount of work, and I can confirm that some work will be necessary. However, the process of reasoning will itself be very interesting. In this chapter I have an opportunity to show you what basic theoretical physics is like. It involves testing an idea by working out carefully what it implies, and then seeing if one can make physical sense of the mathematical conclusions. If one's mathematical ability is sound, and if one's physical insight is also good, then one can discover profound scientific knowledge. It then requires just the right amount of courage to announce the new insight, combined with integrity to make clear what reservations one may still have.

Let us see how we can manage this.

10.1 Einstein's box

Einstein first developed the relation between energy and mass through technical arguments, but having done so he tried to find a simple proof that would be more widely accessible. One argument he gave has become known as 'Einstein's box'. It is not a thorough proof, but it is suggestive. It has a weakness, from a fundamental point of view, in that its starting point is to assume a known relation between the energy and momentum associated with light-waves. This means that it is an argument that is logically based on prior knowledge about the properties of electromagnetic waves. A really fundamental argument would not need that. Nevertheless, let us take a look at the argument at this stage, to acquire a flavour of the ideas.

It can be shown that electromagnetic waves both carry energy and exert forces. An ordinary plane electromagnetic wave has electric and magnetic fields, oscillating in a direction perpendicular to the direction of propagation of the waves. When such a wave impinges on a material object containing charged particles (such as the electrons in the atoms of ordinary substances), the electric field forces the particles of the object to oscillate to and fro in a sideways motion. When they do this, they experience a force from the magnetic field of the wave, pushing them in the direction of travel of the wave. The net result is a force on the object. This force is small in everyday circumstances, but is nonetheless real, and can be measured by sensitive experiments.

For a pulse of light of duration t that delivers energy E when it is absorbed by an object, it is found that the size of the force is $E/(tc)$, which results in net momentum change of the object of $p = E/c$.

With this relation in hand, now consider the following thought experiment.

A box of rest length L sits on a frictionless table. Select an origin of coordinates on the table at the centre of the initial position of the box. The centre of mass of the box therefore lies at $x = 0$. Let M be the mass of each of the two end walls A and B, and suppose the side walls of the box are made of some light elastic material. The total mass of the box is very close to $2M$.

Suppose some molecules on the surface of one end wall A become excited, by thermal agitation, and subsequently emit a pulse of electromagnetic radiation. The pulse has energy E and travels down the box until it is absorbed by the other end wall B of the box. Let us consider the resulting motion of the two ends of the box, in the first instance ignoring any possible relation between mass and energy.

When the pulse is produced, the momentum it will carry to the other end has to be provided, so the molecules that emitted it recoil. This recoil pushes the emitting end wall A, so that the wall acquires a momentum equal and opposite to that of the electromagnetic pulse. This momentum is $p = E/c$. Using the formula 'momentum equals mass times velocity', which is valid at low speed, we deduce that the recoil velocity of the end wall is $v_A = E/(cM)$ to the left. After a time $t = L/c$ the pulse reaches the other end of the box and is absorbed by end wall B. End wall B acquires the momentum (recoiling to the right). At this instant, wall A has already moved through a distance $v_A t = (EL)/(c^2 M)$. Therefore, the left wall has moved but the right wall has not. It follows that the centre of mass of the box is now at

$$x_{\rm cm} = -(EL)/(2Mc^2) \qquad (10.1)$$

where the factor 2 accounts for the fact that we have to average over the mass and position of both ends. In the subsequent motion, the end walls move a little further until forces from the side walls, including damping within the side walls, bring them to a halt, but during all this subsequent motion the centre of mass of the box does not move.[1]

The net result is that owing purely to *internal* dynamics within the system of 'box plus electromagnetic radiation', the centre of mass of the box has moved: see equation (10.1). However, it is a fundamental principle of mechanics that *internal* forces cannot displace the centre of mass of any system. We experience this,

[1] Einstein originally proposed the argument in terms of a rigid box that moves as a whole; but this is an unphysical assumption and is sufficiently incorrect as to make us uneasy that an argument making use of it could not be relied upon. Therefore, here we have avoided it. Einstein himself would of course have been aware that that assumption was not crucial to his argument.

for example, if we take a step forwards in a small rowing boat: the boat lurches backwards, in such a way that the centre of mass of 'human plus boat' does not move. More generally, it can be understood as following from the principle of *conservation of momentum* that we will discuss in the next section. (To shift the centre of mass would require giving it a non-zero velocity for some time, but during such a period the net momentum of the system must be non-zero; that is, changed from its initial value, which breaks the conservation law.)

In order to maintain the strict principle that a centre of mass cannot displace itself by use of internal interactions within an isolated system, we must change the principles of mechanics. Following Einstein, we propose that what happens is that when any object gives up or receives energy E, its mass must also change by some amount m, to be discovered.

Now we work through the argument again, allowing for this. When end wall A releases energy E, we assume its mass changes from M to $M - m$ (with m to be discovered), so its recoil velocity is $v_A = E/(c(M - m))$. After end wall B receives the energy, it has mass $M + m$. At this moment, the two end walls have masses $(M - m)$ and $(M + m)$ and lie at positions $-(L/2 + v_A L/c)$ and $+L/2$ respectively. Therefore the centre of mass of the box is now at

$$\begin{aligned} x_{\text{cm}} &= (M - m)\left(-\frac{L}{2} - \frac{EL}{c^2(M - m)}\right) + (M + m)\frac{L}{2} \\ &= mL - \frac{EL}{c^2} \end{aligned}$$

If we insist the centre of mass cannot be displaced by internal effects, then we must have $x_{\text{cm}} = 0$ so mL must be equal to EL/c^2. Therefore

$$E = mc^2 \tag{10.2}$$

The physical content of this result is that if a material object gives up energy E by emitting electromagnetic waves, then the mass of the object must fall by E/c^2. Conversely, if a mass decrease associated with emission of electromagnetic waves occurs, then the total energy of the waves is mc^2.

We can now immediately argue that the result must be more general, applying to energy in any form, not just electromagnetic waves. For, after a system has given up some energy by electromagnetic radiation, and suffered a mass decrease, one could always restore the energy by some other type of process, such as by heat conduction or collisions with a beam of neutrinos. If the system did not then regain its mass, one could continue in a cycle until all the mass had disappeared but with no other change in the system. All the particles of the box would still be there, but somehow having no mass. This is impossible, so energy and mass must always go together.

In the following sections we will show that the relation (10.2) is indeed general, and one does not need to appeal to properties of electromagnetic waves in order to deduce it. In fact, with hindsight we will find that quoting the relation $p = E/c$ for electromagnetic waves almost amounts to assuming the final result at the outset, so we would certainly prefer a more basic argument if we can produce one. We proceed by first learning about momentum, and then going on to energy. In Relativity, energy and momentum become intimately linked, in a manner similar to the close relationship between time and space.

10.2 Momentum

In passing from kinematics to dynamics, we need to introduce something new beyond the two Main Postulates of Relativity. This is because merely setting up measures of distance and time will not in itself determine what happens when one physical object interacts with another. Having learned from classical physics, most physicists would feel that the most natural next step would be to introduce either some sort of definition of force, or of momentum, or both. There is more than one way to proceed. One could start by saying 'the relativistic momentum of an object is defined to be the product of force and time interval, when a constant force acts for a given time on a given object starting from rest, and I will tell you how to calculate the force in any given situation in a moment.' This method is perfectly logical and correct. However, we prefer arguments based on general notions of symmetry or constancy, if we can find them, and it turns out that we can. Therefore, instead of defining momentum in terms of force, we will define momentum another way, and then define force to be the rate of change of momentum.

The basic insight we need is that of *conservation laws*. A conservation law in physics is a law concerning change: it is a statement that something does not change as time goes on, even during a complicated process where many other things may be changing. For example, there is a conservation law of footballs. This says that as long as no footballs pass into or out of a given spatial region, then the number of footballs in that region is constant, even though they may be being kicked around or thrown or stored. Of course, this conservation law begins to break down when footballs come apart at the seams, or are manufactured from raw materials: it would no longer apply in a football factory for example. Therefore, we do not claim the 'conservation of footballs' to be any sort of fundamental physical law.

We next could formulate a law of 'conservation of electrons', which says the number of electrons in a given region of space is constant as long as none pass across the boundary of the region. This law is upheld under a very wide range of circumstances, much wider than the football law, but it too can break down when electrons and positrons are created or destroyed in high-energy collision processes. Therefore, we go on to guess more abstract laws, such as the law of conservation of electric charge. This is obeyed by almost all physical processes, and is much more widely obeyed than the electron law.

So far we discussed the counting of discrete entities such as particles or electric charge, but we can also consider the possibility that quantities related to motion are conserved. In classical physics, an important conservation law is the law of conservation of momentum (used in the previous section). This says that the sum of all the momenta of a set of objects is constant, as long as the set of objects is not subject to external forces. That is, the objects can pull or push on each other as much as they like, but they will never be able to change their net momentum. Even in the maelstrom of a furnace, if one particle gives up some momentum, then some other particle or particles must pick up exactly that amount of momentum.

The law of conservation of momentum seems to be a basic and profound observation about the universe, so we would like to investigate whether Special Relativity has anything to say about it. Is such a law possible in Special Relativity? It turns out that the answer is 'yes', as long as we define momentum the right way (it will differ from the classical definition). The logic of the argument goes as follows. We propose as a postulate—as a working assumption—that some sort of momentum-like property is conserved, meaning that the sum of it over members of any set of objects will not change when those objects interact, as long as no third parties outside the set are involved. We guess this quantity has something to do with the mass and velocity of each object, so we propose

$$\mathbf{p} \equiv m_0 g[v]\mathbf{v} \tag{10.3}$$

That is, we *define* momentum to be the product of mass and some function (as yet unknown) of velocity. The notation $g[v]$ is shorthand for this unknown function that we want to discover. The square bracket with v inside indicates that we expect the function to depend on v in some way. It might also depend on universal constants such as the speed of light c, but we are proposing that the only way the motion of the particle influences the value of g is through the particle speed v. Is this function perhaps $g[v] = 1$ or $g[v] = 1 + v/c$, or something else? We do not know yet. However, we have built into the definition that the momentum is a vector pointing in the same direction as the velocity. (If one tries to have it point in some other direction, then one soon finds one cannot get a consistent conservation law that way). The zero on the symbol m_0 is to make it clear that this symbol has nothing to do with speed: it is the mass that the object is observed to have when at rest. This is called its *rest mass*.

Having made this working assumption, we will show that Relativity, via its two Main Postulates, determines the function $g[v]$. The way it works is this. If our new 'relativistic momentum' is going to have a conservation law, then the Principle of Relativity says that the conservation law has to apply in all reference frames. It will turn out that this is enough to determine the function! In this sense one may say that the relativistic equation for momentum is 'forced on us'; it is not a separate axiom. The only hypothesis we need to make is that something of the general form (10.3) is conserved. We may as well call that something 'momentum'. It will then require experiments to check whether the hypothesis is right.

Now for the argument. Suppose we have two identical particles moving towards one another with identical speeds. According to our definition (10.3) their momenta are equal and opposite so there is zero total momentum. After the collision, the two momenta must also be equal and opposite, otherwise momentum would not be conserved. (A non-zero total momentum after the collision would be a total momentum that somehow 'appeared from nowhere'—the very thing we are assuming is impossible when we claim that momentum is conserved.) Finally, we shall restrict our analysis to the case of collisions where the rest masses do not change, and the final speeds are the same as the initial speeds. All that happens is a change of direction. We need only argue that collisions of this type are perfectly possible. (They are called 'elastic' collisions, and in fact they are quite common, as any snooker player will tell you.)

So the general sort of collision we have in mind is as shown in Figure 10.1. We assume any sort of speed is possible, and any sort of angle change. Now we notice that for these collisions a nice choice of coordinate axes is always available, to keep the analysis simple, as shown in the figure. The net result is two particles approaching one another along a line oriented at some angle to

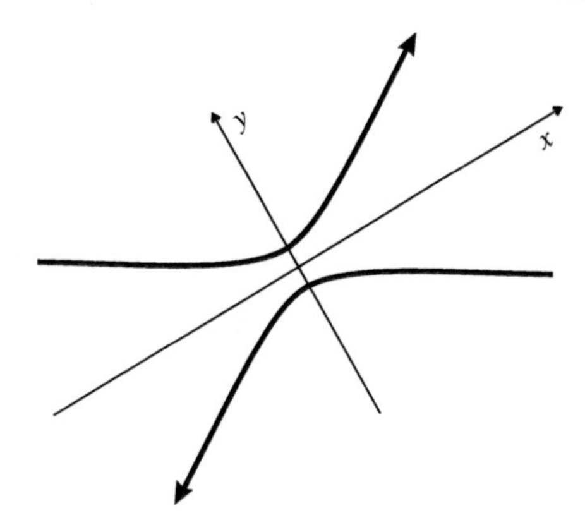

Figure 10.1: A general collision of two identical particles initially moving towards one another with identical speeds. No matter what angles are involved, we can always choose a set of (x, y) axes oriented as shown, relative to the trajectories, in order to simplify the analysis.

the x axis, and leaving along a line making the same angle with the other side of the x axis. Notice that with this choice of axes, the x-component of the momentum of either particle is completely unchanged by the collision, so we have already taken care of conservation of momentum for the x component. It is only the y components (that is, the amount of momentum in the y direction) that we have to worry about.

Now comes the trick. We analyse the collision from two points of view; that is, in two reference frames. First we pick the reference frame moving to the left (along the negative x direction) and keeping pace with the first particle, then we choose the reference frame moving to the right and keeping pace with the second particle: see Figure 10.2. Let the relative speed of these two reference frames be v (of course this speed is related to the speeds of the particles as in Figure 10.1, but we will not need to know what the relation is). In the first (left-going) reference frame the lower particle simply moves up and down at some speed u. From the symmetry of the whole picture, we then know that in the second reference frame the upper particle must move down and up at that same speed u. Let the vertical component of the speed of the other particle be u' in each case (this also

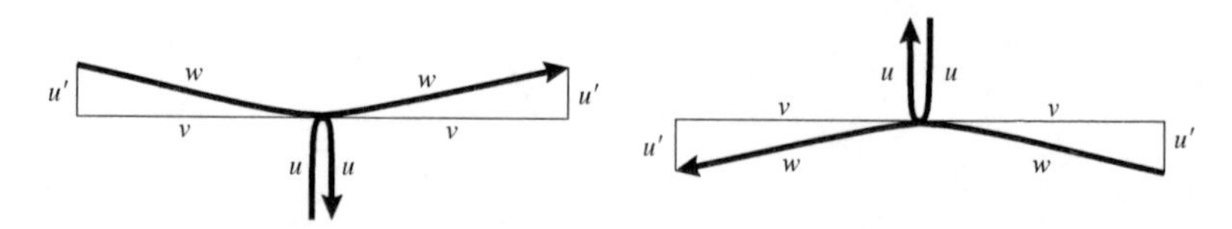

Figure 10.2: The same collision as in Figure 10.1, but viewed from two different reference frames: one moving left keeping pace with the lower particle, and one moving right keeping pace with the upper particle. The relative speed of these two frames is v. From the symmetry of the whole problem one can see that the vertical speed of the upper particle in frame 1 is the same as the vertical speed of the lower particle in frame 2, and so on. w is a total speed and u' and v are its components in the horizontal and vertical directions.

applies in both reference frames, by symmetry). This information is indicated on Figure 10.2.

Now use our definition of momentum, eq. (10.3). In the first reference frame, the momentum of the lower particle is $m_0 g[u]u$ vertically upwards before the collision, and $m_0 g[u]u$ downwards after the collision, so it undergoes a net change of $2m_0 g[u]u$.

The other particle has total momentum $m_0 g[w]w$ before the collision, directed along its velocity $\mathbf{w}$. You can see from the diagram that $\mathbf{w}$ consists of v and u' at right angles, so the total momentum $(m_0 g[w])\mathbf{w}$ is made up of a horizontal momentum component $(m_0 g[w])v$ and a vertical momentum component $(m_0 g[w])u'$. Note that $g[w]$ appears in both of these formulae, not $g[v]$ or $g[u']$. To get the idea, you can think of the combination $(m_0 g[w])$ as an 'adjusted mass' m; then all we are doing is saying that $m\mathbf{w}$ is made of two parts—mv in the horizontal direction, and mu' in the vertical direction.

After the collision, the speed w of the upper particle is unchanged but its direction is changed, such that the horizontal component remains the same but the vertical component reverses. Therefore, the net change is in the vertical direction and is equal to $2m_0 g[w]u'$.

Now we assert conservation of momentum: the change in the momentum of the lower particle equals the change in momentum of the upper one:

$$2m_0 g[u]u = 2m_0 g[w]u'$$

and therefore

$$g[u]u = g[w]u' \tag{10.4}$$

We are trying to find the function g. Can we derive it from equation (10.4)? The answer is 'not yet', because at the moment u appears to be completely unrelated to the other speeds w, and u', and we do not have any equation relating them. However we have not yet made use of the other reference frame. What can it teach us? This is very useful, and if you examine Figure 10.2 it should be clear to you that the vertical speed u' is related to u, simply by a change of reference frame. This is because everything is symmetric: the speed of the upper particle in the right-moving reference frame is the same as the speed of the lower particle in the left-moving reference frame. But we already know what happens to a transverse velocity when we change reference frame: eq. (7.14) says it is reduced by the γ factor, owing to time dilation:

$$u' = \frac{u}{\gamma_v} \tag{10.5}$$

The symbol γ_v is just our old friend γ, but we write γ_v because there are several speeds in play now, all with their own γ factor, and we need to be clear which one we mean. Make sure you are clear that the relative speed of the two reference frames here is v, not $2v$, nor anything else.

Notice that since γ is always greater than or equal to 1, u' comes out *smaller* than u. Therefore, in equation (10.4) we shall find

$$g[w] > g[u]$$

This is saying that because the particle moving at speed w has the smaller *vertical* speed, it needs to have a greater $g[w]$ factor to give it some extra 'oomph' so that it can reverse the momentum of the other particle. In other words, the g function is not going to be simply equal to 1; it will depend on speed in some way.

To find out exactly how, we need to express the three speeds in equation (10.4) in terms of just two speeds. Equation (10.5) tells us how to write u' in terms of v and u, so let us try to do the same for w. By applying Pythagoras' theorem to one of the triangles shown in Figure 10.2, we find

$$w^2 = v^2 + (u')^2 = v^2 + u^2 - u^2v^2/c^2 \tag{10.6}$$

where the second version made use of equation (10.5).

We now have enough information. By substituting (10.5) into (10.4), we find

$$g[w] = \gamma_v \, g[u] \tag{10.7}$$

where w is related to u and v by (10.6). This can be solved for the unknown function g. However, it is tricky to do it by algebraic manipulation. To get a hint, try a simple case. When u is very small, w becomes almost the same as v, and we know that $g[u]$ should go to 1 for small u, in order to produce the classical formula for momentum (simply 'm_0u'). In this simple case (10.7) becomes $g[v] = \gamma_v$. With this hint, we guess that the general solution, for all speeds, is

$$g[v] \stackrel{?}{=} \gamma_v = \frac{1}{\sqrt{1 - v^2/c^2}}$$

This proposal is saying at once:

$$g[w] = \frac{1}{\sqrt{1 - w^2/c^2}} \qquad \text{and} \qquad g[u] = \frac{1}{\sqrt{1 - u^2/c^2}}$$

Having guessed this answer, you can substitute it into (10.7) and verify (if your algebraic skills are sharp) that it works. It is the correct solution, valid for all velocities.

The grand conclusion is

The momentum of a particle of velocity v and rest mass m_0 is

$$p = \frac{m_0 \mathbf{v}}{\sqrt{1 - v^2/c^2}} = \gamma m_0 \mathbf{v} \tag{10.8}$$

This is the formula that we previewed in equation (2.1). You may be relieved to know that this is the hardest bit of algebra you will be shown in the main text of this book.

We have not yet proved that this momentum can be conserved in *all* types of collision process without breaking the Principle of Relativity. However, what we have shown is that if momentum is going to be conserved at all (in all reference frames), then it must have the form (10.8). Our argument also shows that the third Postulate mentioned in the list in Chapter 6 has some hope of being consistent with the other two. To prove that it is consistent with the Main Postulates in all scenarios requires more advanced methods. The following challenge invites you to explore further at this stage, should you wish to.

Challenge. (open-ended). Does the conservation of momentum in collisions between particles of equal rest mass imply such a law for collisions between particles of unequal rest mass? For example, consider the collision of a hydrogen atom H with a hydrogen molecule H_2. If it can be considered as a sequence of pair-wise collisions that each conserve momentum, then momentum must be conserved overall. Can such an argument be made convincingly? Can it be extended to all collisions?

10.2.1 DON'T STOP ME NOW

...
don't stop me now;
Don't stop me
'Cause I'm having a good time, having a good time,
I'm a shooting star leaping through the skies

Like a tiger defying the laws of gravity;
I'm a racing car passing by like Lady Godiva,
I'm gonna go, go, go,
There's no stopping me.

I'm burning through the skies,
Two hundred degrees,
That's why they call me Mister Fahrenheit,
I'm trav'ling at the speed of light,
I wanna make a supersonic man out of you.

Don't stop me now,
I'm having such a good time
. . .

Freddie Mercury © EMI Music Publishing Ltd

Equation (10.8) tells us some interesting things about the way high-speed particles behave. The momentum becomes huge as the speed approaches the speed of light, tending to infinity in the unattainable limit $v \to c$ (unattainable for anything possessing rest mass—the case of light we will look into later). What does a huge momentum mean exactly? Well, we defined momentum to be 'a quantity that is conserved in collisions', so a huge momentum means it would be very hard to stop the particle. If we tried to stop it by throwing slow-moving particles at it as it approaches us, we would need a lot of them. We would need enough that their combined momentum adds up to the big momentum of the oncoming particle. Equally, it would be hard to speed up the particle any further. Perhaps the second fact is not too surprising. It is as if the particle is saying 'please do not ask me to increase my velocity any more: I am pushing up close to the speed of light already.' That makes sense: we know that the light speed is an impassable limit. However, it is quite interesting that the particle also refuses to slow down! By its huge momentum it is saying 'watch out! I'm coming through, don't try to stop me or I'll just knock you down!'

10.3 Energy

Having assumed that momentum is conserved in all types of process, we are free to assert that it is conserved in processes

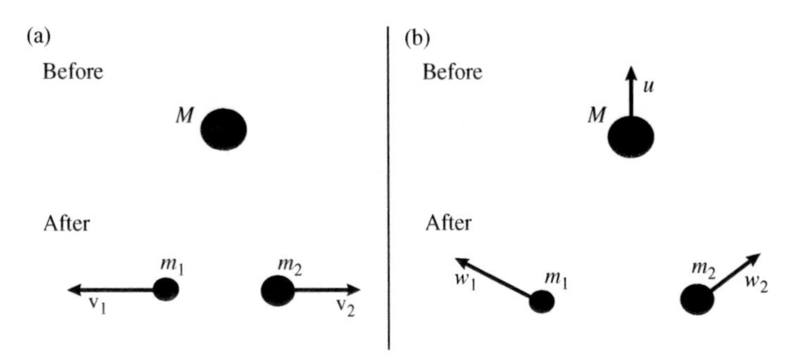

Figure 10.3: An explosion in which a particle breaks up into two pieces. (a) shows the situation before and after the explosion, in the reference frame where the initial particle is at rest. (b) shows the situation in a reference frame moving downwards at speed u relative to the first reference frame.

such as the one illustrated in Figure 10.3. A body or particle of rest mass M splits in two, and the two fragments fly apart. In the reference frame where the initial particle is at rest, there is no momentum before the process, and therefore the total momentum of the fragments is zero. This means they must fly off in opposite directions—shown horizontal in figure 10.3a.

Next consider a reference frame moving at speed u relative to the first, in the vertical direction. Let the speeds of the fragments be w_1, w_2 in this reference frame. Clearly the speed of the initial particle in this new reference frame is u, and this is also the vertical component of speed for both the fragments after the collision. To see this, imagine that the particles are moving along a horizontal wire; in the new reference frame the wire is still horizontal and it is moving at speed u upwards. It follows that the total momentum in the vertical direction is $M\gamma(u)u$ before the collision, and $m_1\gamma(w_1)u + m_2\gamma(w_2)u$ after the collision. Therefore, conservation of momentum gives

$$M\gamma(u)u = m_1\gamma(w_1)u + m_2\gamma(w_2)u$$

Therefore

$$M\gamma(u) = m_1\gamma(w_1) + m_2\gamma(w_2) \tag{10.9}$$

This is a marvellous and remarkable equation. We are free to choose u as small as we like, and in the limit of u going to zero we have an equation for the situation in the rest frame of the initial object:

$$M = m_1 \gamma(v_1) + m_2 \gamma(v_2)$$

where v_1, v_2 are the speeds of the fragments in this frame. For brevity, let us write this as

$$M = m_1 \gamma_1 + m_2 \gamma_2 \tag{10.10}$$

This equation says that the rest mass of the initial object is larger than the sum of the rest masses of the products! Consider, for example, the case of a 1-kilogram mass that splits into fragments moving at $v_1 = v_2 = 0.999998\,c$, which gives $\gamma_1 = \gamma_2 = 500$. In this case, according to equation (10.10), the fragments must have a rest mass of one gram each (since one kilogram is 1,000 grams). Therefore, before the explosion there was 1 kilogram of material and after it there is material which, if it were brought to rest, would exhibit a total mass of 2 grams. So 998 grams of rest mass has disappeared from the world. Where has it gone? I know you are thinking 'it has been converted into energy', but before you jump to that (correct) conclusion I want to invite you to ask yourself what you mean by that. The essential idea is that the loss of rest mass occurs whenever the γ values are greater than 1, which means whenever the products are *moving*. So we have a conversion between rest mass and *motion*. The word 'energy' is a useful shorthand for discussing either a property of moving things (called *kinetic energy*) or a property of things that have the potential to cause motion (called *potential energy* or *internal energy*). We *define* the 'energy' to be 'a quantity associated with motion, or the potential to cause motion, that is conserved when things interact with one another.' Equation (10.10) is expressing just such a conservation. On the left we have a property of the object before the explosion, and on the right we have properties of the fragments after the explosion, which are clearly associated with their motion. If we write the equation this way:

$$M = m_1 + (\gamma_1 - 1)m_1 + m_2 + (\gamma_2 - 1)m_2 \tag{10.11}$$

then we can interpret the terms in the equation as follows:

Property	Quantifies
M	The propensity of an object at rest to cause motion if it splits up
$(\gamma_1 - 1)m_1$	Motional energy (it is zero when $v_1 = 0$)
$(\gamma_2 - 1)m_2$	Motional energy (it is zero when $v_2 = 0$)
m_1	The propensity to cause further motion should this fragment further divide
m_2	The propensity to cause further motion should this fragment further divide

We are discovering a *dual role* for the property that we have been calling 'mass'. We introduced mass in the first instance as a measure of *inertia*—the tendency of a body to resist being accelerated when other bodies hit it. This was its role when it entered into the formula for momentum. However, we now have a new role for the same property: it also quantifies the propensity of a body to produce motion when the body splits up into fragments. This new role is the very thing we normally call energy. Therefore, mass and energy always go together. They are, in short, just different human words for the same physical property.

'But what about conservation of mass?' you say. Well, it looks as though mass, in the traditional sense of rest mass, is *not* conserved. Equation (10.10) says that something else *is* conserved: namely, the quantity γm. For, on the left we have the rest mass of something which is not moving, so it has $\gamma = 1$, while on the right we have the sum of γm for all the particles contributing to the collision.

Approximate expression for γ

The definition of γ can be written

$$\gamma = \frac{1}{\sqrt{1-x}}$$

where $x = v^2/c^2$, and we are interested in the case where x is small compared to 1. Taking the square and then multiplying both sides by $(1-x)$ gives

$$(1-x)\gamma^2 = 1 \qquad \Rightarrow \quad \gamma^2 = 1 + x\gamma^2$$

This is exact. At $x = 0$ we have $\gamma = 1$, and for small x, γ only departs from 1 by a small amount. Therefore, on the right-hand side of the equation we can put $\gamma \simeq 1$ and then the equation gives

$$\gamma^2 \simeq 1 + x \tag{10.12}$$

Next use the fact that for any x,

$$\left(1+\tfrac{1}{2}x\right)^2 = \left(1+\frac{x}{2}\right)\left(1+\frac{x}{2}\right) = 1 + \frac{x}{2} + \frac{x}{2} + \frac{x^2}{4} = 1 + x + \tfrac{1}{4}x^2$$

and when x is small compared to 1, the x^2 term is small compared to x and can be dropped:

$$\left(1+\tfrac{1}{2}x\right)^2 \simeq 1 + x \qquad \text{for } x \text{ much smaller than 1}$$

For example, $x = 0.2$ gives $1.01^2 = 1.0201 \simeq 1.02$. Applying this to (10.12) gives

$$\gamma^2 \simeq \left(1+\tfrac{1}{2}x\right)^2 \qquad \Rightarrow \quad \gamma \simeq 1 + \tfrac{1}{2}x = 1 + \frac{1}{2}\frac{v^2}{c^2}$$

which is equation (10.13).

Now look again at equation (10.9). It expresses the same idea, but now for an example where everything is moving, the initial speed being u and the final speeds w_1, w_2.

We now have the essential idea of the equivalence of mass and energy. However, we are missing the 'conversion factor' c^2 in the formula $E = mc^2$. To discover this factor we need to examine the terms $(\gamma - 1)m$ that we called 'motional energy' in the interpretive summary above. At low speeds the formula for γ can be written

$$\gamma \simeq 1 + \frac{1}{2}\frac{v^2}{c^2} \quad (10.13)$$

The proof of this is given in the box, and you can also check it using a calculator. For example, at $v/c = 9/41 \simeq 0.22$ the exact result is $\gamma = 1.025$, while the approximate formula gives 1.0241. Therefore, at low speeds,

$$(\gamma - 1)m \simeq \frac{1}{2}mv^2/c^2$$

At low speeds the energy associated with motion, called kinetic energy, is given by a standard formula:

$$\text{kinetic energy at low speed} = \frac{1}{2}mv^2$$

where m is the rest mass. For example, a 1-kilogram mass moving at 1 metre per second has a kinetic energy of 0.5 joules. This kinetic energy is c^2 larger than the value we just found for $(\gamma - 1)m$. Therefore at low speeds, $(\gamma - 1)m$ understates the kinetic energy possessed by the moving fragments by a factor c^2. Therefore, to interpret equation (10.11) correctly, we should multiply it by c^2, and then each term represents an amount of energy:

$$Mc^2 = m_1c^2 + (\gamma_1 - 1)m_1c^2 + m_2c^2 + (\gamma_2 - 1)m_2c^2$$

The physical interpretation is now:

Property	Interpretation
Mc^2	internal energy
$(\gamma_1 - 1)m_1c^2$	kinetic energy
$(\gamma_2 - 1)m_2c^2$	kinetic energy
m_1c^2	internal energy
m_2c^2	internal energy

To summarize:

$$\left.\begin{array}{l}\text{Conservation of momentum,}\\ \text{Principle of Relativity}\end{array}\right\} \Rightarrow \begin{array}{l}\text{Conservation of `adjusted}\\ \text{mass' } \gamma m_0\end{array}$$

$$\text{physical interpretation} \Rightarrow \gamma m_0 c^2 \text{ is energy}$$

where in the last line we added the internal energy and kinetic energy together, to make the total energy:

$$E = \gamma m_0 c^2 \qquad (10.14)$$

This is the famous equation, $E = mc^2$. When people quote the simpler form they either have in mind a definition $m = \gamma m_0$, or they just want to quote the result for the case of a stationary object, where $\gamma = 1$. We used m_0 in this summary in order to emphasize that the mass appearing in the formula we have given is rest mass (the mass a body has when it is not moving). The phrase 'conservation of adjusted mass' means that if you add up all the γm_0 values for particles before some process (each particle having the γ associated with its speed), and then again after the process, the individual terms in the sum may change, but the sum total will be the same. If, in carrying out the sum, we multiply every term by the same factor c^2, then the total will also be the same before and after the process, so we have a conservation law for $\gamma m_0 c^2$. This is called the 'conservation of energy'.

10.3.1 DISCUSSION

It may be that you still feel slightly uneasy about the status of the result. It may appear that we did not so much prove the equivalence of mass and energy as just claim that it is true. To help with this, I shall summarize the argument. We started with conservation of momentum. We found that if momentum is to be conserved in all reference frames, then it has to be related to mass and velocity in a certain way, $\gamma m_0 \mathbf{v}$. We then found that if momentum is conserved in all reference frames, then so is the quantity γm_0. This is a quantity that is conserved, and that can be converted from one form to another, including from a stored form to a motional form, and that reproduces the behaviour of classical energy at low velocities as long as it is multiplied by c^2. Therefore, $\gamma m_0 c^2$ has the complete behaviour of the quantity we normally call energy. But if it looks like energy and it smells like energy, it *is* energy.

Is it right to call γm_0 'mass'?

Many authors like to refer to the quantity γm_0 as 'mass'. Others prefer not to call it that, and instead reserve the word 'mass' to refer only to rest mass. Calling γm_0 'mass' means that one may speak of 'an increase of mass with velocity', which provides a good intuition about what happens when particles move fast: it becomes harder and harder to accelerate them further. However, in Relativity the ratio between force and acceleration is not simply equal to γm_0, so one should keep in mind that γm_0 is not altogether the same as what we normally call mass. I prefer not to call it mass because we already have a perfectly good name for it: energy! (up to a multiplying constant). In this book the symbol m *always* refers to rest mass, and I never describe γm_0 simply as 'mass'. This chapter is the only place where I come close to doing that, when I make occasional use of the phrase 'adjusted mass'. In Relativity the most important mass-like quantity is the rest mass, because its value does not depend on a reference frame (whereas energy and momentum do).

The point here is, of course, not that we are merely naming something. Rather, we have uncovered a profound connection: a connection between mass and energy. Mass is the *inertial* property—a propensity to resist acceleration—and it enters into the formula for momentum. Energy is the *invigorating* property—a propensity to cause motion. According to Special Relativity, these two propensities are so intimately linked that they are one and the same, up to a universal constant multiplying factor.

> *Associated with the inertial property called 'mass' there is an energy. Equally, energy is always accompanied by inertia.*

This is one of the most profound insights achieved by Relativity theory.

Full conviction will come once you have seen the idea at work in enough examples, and, crucially, once we have checked that the predictions match experimental observations. Let me assure you that they do! Although we have considered only one type of collision for the relativistic energy calculation above, more complicated collisions can be broken down into simpler ones, at least for the purpose of keeping account of the energy. For example, in a collision of the form $a + b + d + e \rightarrow f + g + h$ one can imagine that first a and b come together, then the joint object collides with d, and the new joint object collides with e. The total result is an object that proceeds to split into two products, one of which splits again. Thus, repeated application of equation (10.9) is sufficient to track the energy conservation in any collision process.

Here are some examples.

- In an explosion of an object initially at rest, 1 gigajoule of energy is released in the form of kinetic energy of the products. What is the change in rest mass (that is, the difference between the rest mass of the initial object and the sum of the rest masses of the fragments)? Answer: $(10^9\,\mathrm{J})/c^2 = 11$ micrograms. (A microgram is one millionth of a gram.)

- Suppose a spring obeying Hooke's law supports a car of mass 1,000 kg, being compressed from its natural length by $L = 1$ metre. What is the mass increase of the spring? Answer: The spring constant is $k = mg/L = 9800$ N/m, and the stored energy in the spring is $(1/2)kL^2 = 4900$ joules. The rest mass of the spring is therefore increased by $E/c^2 \simeq 0.05$ nanograms (5×10^{-14} kg).
- In a kilogram of helium gas at ordinary temperatures, there are 1.5×10^{26} helium atoms. If such a gas is heated, the total kinetic energy of the atoms increases with temperature by approximately 6, 200 joules per degree of temperature change (measured in either Celcius or Kelvin). If the gas is heated from room temperature (20°C) to 'white hot' (5520°C), what is the change in rest mass of the gas? Answer: $(6,200 \times 5,500)/c^2 = 0.38$ microgram.

In the last example, the energy associated with heating of the gas goes to *kinetic* energy of the molecules, but by the 'rest mass of the gas' we mean the propensity of the gas as a whole to resist being accelerated by forces applied to the walls of the chamber containing it. No individual molecule changes its rest mass; it simply moves faster. However, these faster-moving molecules carry more momentum, and this means that they exert a greater force on the walls of the chamber. If an external force is applied to the chamber, to accelerate it, it will be found that the inertial mass of the gas as a whole has increased: it exhibits more 'reluctance' to be accelerated. It is very remarkable that one can increase the mass of a gas (or anything else) merely by heating it. The mass increase of 0.38 microgram might appear small, but that depends what you compare it with. It is not small on an atomic scale: it is equal to the rest mass of some 10^{17} helium atoms.

It is a common misconception that the relation $E = mc^2$ applies only to nuclear effects such as nuclear fission and fusion. In fact, it applies to all energy effects, and we have just considered some examples. Much of the energy we deal with in everyday life comes from chemical reactions (think of eating a sandwich, or driving a car). The energy of these reactions is also associated with mass.

It is just that the energies are small, so the mass changes are not so noticeable. However, the formula presents us with the stunning fact that vast amounts of energy are 'locked up' in ordinary matter, so that if the whole of this energy could be made available for other purposes, then only a small amount of matter would be needed to supply all our energy needs.

In nuclear fission, a heavy nucleus such as uranium 235 splits when a neutron hits it, the two halves then each undergo further splits, and so on, until stable products such as niobium (Nb) and praseodymium (Pr) are formed. The whole sequence may be summarized thus:

$$^{235}\mathrm{U} + \mathrm{n} \rightarrow \cdots \rightarrow {}^{93}\mathrm{Nb} + {}^{141}\mathrm{Pr} + 2\mathrm{n}$$

The rest masses of all the relevant players are listed in the table. You can see that the total rest mass of the products is smaller than the rest mass of the reactants by 0.221 atomic units. So the proportion of the initial rest mass that is converted into kinetic energy is about 0.1%. This is a small proportion, but it is huge compared to the proportion that is converted in chemical reactions such as the burning of coal, gas, or petrol. However, nuclear fission is not a good long-term prospect because reserves of uranium are limited, and the by-products of the process include highly toxic radioactive waste. Nuclear fusion, on the other hand, is harder to control but is a much better long-term prospect. We shall examine it in a moment.

The ultimate way to release energy is to employ a black hole or some antimatter. Both approaches are ridiculously impracticable to us, but are allowed in principle by the laws of physics. Perhaps far-future human engineers (or present alien ones if there are any) may surmount the difficulties. If you have a black hole nearby you should not approach it, but if it is small you can collect the Hawking radiation it emits, and keep it fed by dropping things into it. If it is too large to emit much Hawking radiation, then you may send a power station into orbit around it. The power station works by lowering objects slowly into the black hole on a long

Table 10.1 The rest masses of some nuclei and particles involved in fission and fusion reactors. The atomic unit of mass is $m_{au} = 9.109382 \times 10^{-31}$ kg, and $m_{au}c^2 = 1.4924178 \times 10^{-10}$ joules.

Nucleus or particle	Mass in atomic units
Uranium ^{235}U	235.04392
Praseodymium ^{141}Pr	140.90765
Niobium ^{93}Nb	92.90638
Helium ^{4}He	4.0026025
Tritium (^{3}H)	3.0160493
Deuterium (^{2}H)	2.0141018
Neutron (n)	1.0086649
Proton (p)	1.0072765

and extremely strong rope, and using the rope to turn a turbine at the power station. By the marvels of General Relativity, the mass energy of the lowered object is gathered at the turbine, and none at all passes to the black hole! You can obtain 100% energy conversion in this way, but the rope needs to have a tensile strength far exceeding that of any known material.

The antimatter method is based on the fact that when antimatter is brought into contact with ordinary matter, the resulting explosion liberates 100% of the mass-energy of both. However, to make use of this for energy production, first you would need a ready supply of antimatter. Antimatter can be manufactured, but it is almost impossible to avoid manufacturing an equal amount of matter at the same time. Therefore, when the energy is extracted (by matter–antimatter annihilation) you merely get back the energy initially invested in the manufacture step. This could still be useful for making fuel or a bomb, though an extremely volatile one. It is within the bounds of physical possibility to imagine a process that carries out a sequence of reactions resulting in a net conversion of some kinetic energy into mass-energy of positrons, without forming an equal number of electrons. If those positrons were made efficiently, then if they subsequently were made to annihilate with ambient electrons, the net result would be the release of the electron mass-energy.

Both these methods are currently impracticable to us, but they are related to real processes that have been measured. There is strong evidence that a process closely related to the 'black hole power station' is taking place in reality near many black holes. There is no rope or power station, but dust particles falling towards the black hole encounter a swirling disc-shaped cloud of matter already in orbit. As they descend through the cloud, in a long spiraling orbit, their motion is inhibited by collisions with other particles in the cloud; this plays the role of the 'rope'. About 6% of the mass-energy of the infalling particles is released into the heating of the disc and electromagnetic radiation, before they finally exit the cloud and are swallowed by the hole, which acquires the remaining 94%.

In particle accelerator experiments, positron production is now routinely carried out in man-made positron sources, but electrons are formed at the same time (and in terms of energy investment it is hopelessly inefficient). The much more rare process of forming matter and antimatter in differing proportions has been observed, but it tends to favour the amount of matter, so is again important fundamental physics but irrelevant to energy production.

Mathematical Challenge. This challenge, should you accept it, will allow you to confirm the overall consistency of the ideas in slightly greater generality than we have so far. The method is to apply the relativistic equations for velocity and momentum in a further example. Consider the break-up illustrated in Figure 10.3. The figure shows the situation in the reference frame where the total momentum is zero, and in a reference moving vertically with respect to the first. Repeat the calculation, but now for a reference frame moving at speed u horizontally with respect to the first; that is, along the line of motion of the particles. The final speeds of the particles in the new reference frame are given by the relativistic addition of velocities equation (7.11) (with appropriate choice of signs). Prove that

$$\gamma(w_1)m_1w_1 + \gamma(w_2)m_2w_2 = \gamma(u)\left(\gamma(v_1)m_1 + \gamma(v_2)m_2\right)u$$

where the speeds w_1 and w_2 are in the same direction, and thus prove again that if momentum is conserved then M must be as given by equation (10.10). (Method: first look into the γ factors, and prove that

$$\gamma(w) = \gamma(u)\gamma(v)(1 + uv/c^2)$$

when $w = (u + v)/(1 + uv/c^2)$. Apply this to both w_1 and w_2, keeping in mind that v_1 and v_2 are in opposite directions. Then tackle the formula you are asked to prove. Two terms can be cancelled by arguing from momentum conservation in the first frame, as long as the momentum of any particle is given by the relativistic momentum equation (10.8).)

10.3.2 KINETIC ENERGY, BINDING ENERGY

We have already used the term 'kinetic energy' for the energy associated with motion. The formal definition is

$$K \equiv E - m_0 c^2 \tag{10.15}$$

where m_0 is the rest mass of the object under consideration. This means that we may say the total energy of any body is the sum of its rest energy and its kinetic energy.

It often happens that the rest mass of a physical object is smaller than the sum of the rest masses of the things of which it is made. This happens whenever it would require some energy input to split the object. In this case the set of constituents is said to be 'bound', and the energy required to split them up is called the 'binding energy'. The idea of 'binding energy' is already present in classical physics; Relativity introduces the fact that the binding energy is associated with a reduction in rest mass. For example, the rest mass of a water molecule is very slightly smaller than the sum of the rest masses of two hydrogen atoms and one oxygen atom. The rest mass of any atom is smaller than the sum of the rest masses of its electrons and nucleons.

Binding energy is most noticeable in nuclear physics. The helium-4 nucleus is made of two protons and two neutrons, and its rest mass is 4.0026 atomic units. However, the rest mass of a proton is 1.00728 atomic units, and the rest mass of a neutron is 1.00866 atomic units. Therefore, a helium-4 nucleus is about 0.029 atomic units lighter than the sum of the rest masses of its constituents. This is approaching 1% of the total. If the constituents are provided, then fusing them together will result in the binding energy being released in the form of kinetic energy of the final nucleus. This results in a very fast-moving helium nucleus, and if it happens in a dense plasma then the result is a hot maelstrom. This is what goes on in the heart of the Sun. In prototype fusion reactors on Earth the aim is to achieve nuclear fusion of deuterium and tritium in controlled conditions, making helium plus a neutron plus plenty of kinetic energy. The desired reaction is

$$\mathrm{D} + \mathrm{T} \rightarrow \mathrm{He} + \mathrm{n}$$

for which the mass deficit is 0.0189 atomic units, so generating about 3×10^{-12} joules of energy per reaction. Currently the leading experiments achieve around 10 megawatts of fusion power, representing the fusion of about 30 micrograms of tritium and deuterium per second. The kinetic energy can be used to heat water to drive turbines to generate electricity. Although the technological challenge is great, this is a serious long-term contender for generating large amounts of power without the emission of greenhouse gases, thus easing the pressure placed by humanity on Earth's resources. The theory of Relativity did not invent fission or fusion, of course, but it helps us to understand how it can be that such large amounts of energy can be hidden inside such apparently unprepossessing raw material.

10.3.3 MAKING NEW PARTICLES

We began this chapter with a discussion of 'Einstein's box', and quoted a formula $E = p/c$ for a pulse of light, coming from the theory of electromagnetism and confirmed by experimental observations. In the subsequent discussion of momentum and energy we have mentioned only material particles possessing rest mass. The formulae $\mathbf{p} = \gamma m_0 \mathbf{v}$ and $E = \gamma m_0 c^2$ cannot be applied to light because the γ factor goes to infinity and, as we shall see in a moment, the rest mass is zero. However, by combining these formulae we can obtain two useful equations:

$$E^2 - p^2c^2 = m_0^2c^4 \quad (10.16)$$

and

$$p = Ev/c^2 \quad (10.17)$$

(You can now prove these to your own satisfaction.) The results (10.16) and (10.17), expressing relationships between energy and momentum, do not mention γ, and they still make sense when v is equal to c. Imagine accelerating a particle to higher and higher speeds. Such a particle will have more and more energy and more and more momentum, so both the terms on the left-hand side of (10.16) become very much larger than the right-hand side m_0c^4. Also, equation (10.17) shows that in the limit $v \to c$ we have $p \to E/c$. It makes sense, therefore, to expect that a physical entity that achieved the limit $v = c$—for example, a pulse of light—would have $p = E/c$ and therefore (by (10.16)) zero rest mass. This is consistent with what we learned from electromagnetic theory, and it is what is observed experimentally.

Perhaps the most striking example of the equivalence of mass and energy is when particles of zero rest mass interact to create pairs of particles with rest mass. An example is a gamma-ray photon moving in the intense electric field close to an atomic nucleus: the photon can give up its energy to the creation of a particle–anti-particle pair such as an electron and a positron. These two particles then move apart and go on their way and 'have a life'. This is sometimes called 'the creation of mass from energy', but in view of the fact that mass and energy are the same thing, it is better to call it 'the conversion of energy from an electromagnetic form into a material form'. The reverse process also happens. It is called annihilation: a particle and anti-particle pair come together, and all their energy is converted into gamma-ray photons, leaving no rest mass at all.

In particle colliders—the vast machines beloved of particle physicists—pairs of high-energy particles are brought together, and in the resulting collision there is an extremely high concentration of energy into a very small volume. According to the equivalence of mass and energy, in that tiny furnace new particles can be formed. When the experiments are done, new particles are indeed formed; but not just any old particles. Only a fairly modest number of different particles—electrons and quarks and a few others, or composites of those—are observed. In other words it is not possible to give birth to a genuinely novel fundamental particle, in the sense of one at some arbitrary combination of mass and electric charge and spin and other properties. The modern understanding of this is to say that the particles are energetic excitations of fundamental 'fields' that are already there in a quiescent form before the experiment is carried out. Performing a collider experiment is like jabbing a pin into the fabric of spacetime and watching to see what will come out. The particles that emerge tell us about the fundamental fields which appear to be what spacetime is made of.

10.3.4 MASS WITHOUT REST MASS

Suppose I have a pair of mirrors, both at rest, such that the mirrors and their support have a total rest mass m_0. The rest energy is therefore $m_0 c^2$. Now suppose I put some light of energy W in between the mirrors, bouncing to and fro. Now, the object in my possession ('mirrors plus light') has total energy $W + m_0 c^2$. This object is not itself moving: it has moving 'inner workings' (the light), but its centre of mass is not in motion. Therefore, the energy $W + m_0 c^2$ must be its rest energy, and we deduce that its rest mass is $(W + m_0 c^2)/c^2 = W/c^2 + m_0$. That is interesting: it became heavier, even though the stuff I put in had zero rest mass.

Now suppose the mirrors have very little mass, and I put a *lot* of light in. Then the total energy of the object can be mostly carried by the light. The W/c^2 part could dominate, while the m_0 part contributed by the mirrors could be almost negligible by comparison. This raises the intriguing idea that an object could have a substantial rest mass, even though most of the stuff it is made of has no rest mass. So we ask, could the particles (protons, electrons, and so on) of which the universe is made be themselves made from stuff with no rest mass? Does this explain the origin of mass?

This idea has been eagerly pursued by particle physicists. As things stand, it does look as though a lot of the mass of the universe can be understood as arising from interaction energies between particles which themselves have only very little rest mass. We cannot explain all of mass in this way, but we can explain most of it.

Such investigations lead us tantalizingly close to the feeling that we are obtaining a glimpse of the ultimate workings of the universe. If we can derive the mass from the interaction energies, then we need only understand the interactions. The mass will take care of itself, so we have one less thing to explain. So far we have learned that the universe is not quite as simple as that, but with Special Relativity and other advances of physics, old and new, we are certainly on the right track.

10.4 The photon rocket

We can now make a serious study of the possibilities of interstellar travel using rocket technology.

Jet engines and rocket engines both work by a similar principle: material is ejected at high velocity out of the back of the engine, and by conservation of momentum the aircraft or rocket from which the matter is ejected must receive an increase in momentum in the forward direction. One can also understand this in terms of forces: there is a pressure force on the front wall of the chamber from which the matter is ejected. The main difference between a jet and a rocket is that a jet engine sucks in matter (air) from its surrounding and ejects it at higher speed, while a rocket has to carry all its own material with it so that it can operate in the near vacuum of space.

One of the technical aims in the design of rocket engines is to get the most gain in momentum for a given expenditure of rocket fuel. This means that one wants the emitted particles to have the highest possible ratio p/E. Equation (10.17) says that $p/E = v/c^2$, so we want a high speed v for the ejected particles. The highest available is $v = c$, so we conclude that the best one can do is to build a 'photon rocket'. This type of rocket drives itself along by shining powerful searchlights or laser beams out the back. Such 'engines' convert rest mass energy of the rocket fuel into photons (if it is not already in that form), and emit them out of the back of the rocket. Our current technological capabilities are, unfortunately, far from what they would need to be to make a viable rocket this way. (A searchlight consuming 1 gigawatt of power—the output of a modest power station—would only produce only 3 newtons of force; and using a laser is no better, because lasers are quite inefficient.) However, for an interstellar journey where conserving fuel is of paramount importance, it would be a waste to throw anything except particles with $E = pc$ out of the back of a rocket, so to explore the limits of what might be possible in the future, the 'photon rocket' concept is a good way to proceed.

Let m be the rest mass of the rocket. This is continually falling owing to ejection of the energy given to the photons. After some time the rocket has speed v and rest mass m, having emitted radiation now travelling in the opposite direction and carrying energy E_r, momentum E_r/c. By conservation of energy, the initial energy of the rocket, $m_i c^2$, (where m_i is its initial rest mass) must equal the net energy of rocket and photons at any later time:

$$m_i c^2 = \gamma m c^2 + E_r$$

Also, the momentum of the rocket must be equal in size (and opposite in direction) to the momentum of the photons, so

$$\gamma m v = E_r/c$$

From the second equation we have $E_r = \gamma m v c$. Substituting this in the first gives

$$\begin{aligned} m_i c^2 &= \gamma m c^2 + \gamma m v c \\ \Rightarrow \qquad m_i &= \gamma m(1 + v/c) \end{aligned}$$

and therefore

$$\frac{m}{m_i} = \sqrt{\frac{1 - v/c}{1 + v/c}} \tag{10.18}$$

(To check the algebra to derive this, it is helpful to recall that $(1 - v^2/c^2) = (1 - v/c)(1 + v/c)$). For example, suppose we would like the rocket to attain a speed fast enough to make the gamma factor $\gamma = 10$, so that the journey between the stars can benefit from a useful time dilation factor. This value requires $v \simeq 0.995c$, and we learn from eq. (10.18) that the ratio of final mass to initial mass of the rocket would be about 0.05. That is, 95% of the total mass of the rocket has to be given over to fuel, leaving 5% for the engines, living quarters and support structure. This is a small proportion, but it seems feasible.

However, we have not yet taken into consideration that we might want the rocket to slow down at its destination—and we might even want it to come back! Suppose that the rocket 'coasts' at $v = 0.995c$ for the greater part of its journey, without firing the engine, in order to conserve fuel. To bring the rocket to a stop at the destination, the sequence is just as for the initial launch, except that the engine has to point in the direction of motion, so as to act as a brake. To analyse this, adopt one of the standard tricks of relativistic arguments: change reference frame! We choose the reference frame that moves along with the rocket at its coasting velocity. In this reference frame, the coasting rocket begins at rest, and the deceleration sequence is an acceleration from zero velocity up to v, so the same formula (10.18) applies. We deduce that the fuel expenditure to stop the rocket is such that the rest mass of the stopped rocket is 5% of the rest mass of the coasting rocket.

It follows that the ratio of rest mass at the outset of the journey (say, in Earth orbit) to the rest mass at the destination (say, orbiting a planet of some other star), is $0.05 \times 0.05 = 0.0025$. This is beginning to look problematic.

If we can refuel at the destination, then it is just feasible to imagine coming back again. If we cannot then the mass ratio for a full round-trip, including all starts and stops, is $0.0025^2 \simeq 0.000006$. This proportion is so small that it does not appear to be a feasible engineering proposition.

The conclusion of all this is that rocket propulsion on its own does not appear to be a good method for travelling around the galaxy: the fuel consumption is prohibitive unless we are prepared to go more slowly, but that would require centuries to pass *en route*. One is therefore forced to explore alternatives. These include providing impulse from an Earth-based laser beam, scooping up interstellar material *en route* (in the manner of a jet engine), and sailing on the weak solar wind of particles emitted by the Sun.

11

Conclusion

We have come to the end of our explorations. If you now recall the introduction, or reread it, I hope you will find that the observations made there about life on board a fast rocket make sense. Or, I hope you will feel that the arrival of muons at sea level, after travelling for many times their normal lifetime, is no longer surprising, but rather, just what we must expect.

I once had the pleasure of visiting the Palais de la Découverte in Paris. This is one of the most fascinating science museums I know. In a room at the top of the building I came across an exhibit sitting quietly on a table, showing something with which I was familiar but had never seen. It was a cloud chamber (*chambre à brouillard*)—a glass box about thirty cm wide, containing a vapour just on the point of condensing into liquid. Whenever a fast-moving charged particle moved through the chamber, it would leave a trail of tiny droplets of moisture (like the 'vapour trails' left by jet aircraft in the sky, but much much smaller). These little trails would soon evaporate again, but then another would appear somewhere else as the next particle came into the chamber. It was late in the day and I had the exhibit to myself for a few minutes. It was brightly lit, with a slight cloudiness as the vapour was continually on the verge of condensing, with these little trails of droplets appearing and disappearing all the time. In its bright cloudiness interspersed by the dagger-like trails it was a very beautiful object. But what made the experience intensely moving for me was the fact that I knew what these little arrivals were, and I knew their significance to the experimental study of Relativity. I hope you have guessed it: the charged particles arriving in the chamber were atmospheric muons. I marvelled both at the visual spectacle, and at the underlying physics, and at the achievement of the people who had first understood it. An experience like this is akin to hearing Mozart's clarinet quintet or watching a piece of football poetry by a great player. It is not an experience that anyone could have without the hard work and understanding that is required in order to appreciate it, but that would be true of music and football, and poetry too. Such things fulfil a yearning that is deep in all of us. One that is different from—and more precious than, luxury and comfort.

The journey we followed in this book set off from muons and light, and allowed us to understand those things. But more importantly, it revealed a whole landscape of profound and powerful ideas, such as the unity of spacetime, the non-uniformity of time, and the equivalence of mass and energy. You have been invited to think through precisely what that equivalence means, and how it illuminates what goes on in particle reactions, such as in the fusion processes in the Sun, or indeed in ordinary circumstances such as a burning log fire. We also glimpsed the compelling idea that much of the mass of ordinary objects may come from the interaction energies of the fundamental particles (quarks) inside them.

We took two modest steps towards Einstein's greater work: the theory of General Relativity. The first step was the whole spacetime approach; the second step was the connection between uniform or inertial motion and the accumulation of proper time. We learned it from the twin paradox, which is certainly the most important of the paradoxes explored in Chapter 8.

The type of reasoning that Einstein brought to Special Relativity has since been carried further, and to great effect. It is a method that focuses attention on what things *stay the same* when there is a change in perspective. In the case of Special Relativity, the first postulate concerns physical behaviour that stays the same (irrespective of shared uniform motion), and the second postulate concerns the speed of light that stays the same (irrespective of motion of the source). This idea of 'constancy' or 'invariance' under a change is now called a 'symmetry principle'—the single most important theme in the development of physics throughout the twentieth century. In fact, 'relativity' is arguably not the best name for Einstein's theory. It appears that it was Max Planck who introduced the name in 1906, but Einstein had his reservations about it and is quoted as saying:[1] 'Now to the term *relativity theory*. I admit that it is unfortunate, and has given occasion to philosophical misunderstandings.'

A better name might be 'the theory of invariants'—a name that Einstein preferred. When professional physicists use relativity theory in practice, the most important quantities are things like proper time and rest mass. These are absolute properties of spacetime and the particles that inhabit it. They do not depend on any particular choice of reference frame, and they are said to be 'invariant'.

I know that for many readers of this book, the mathematical manipulations, especially in the last chapter, will have involved steps that you had to take on trust. However, I hope that you preferred to see them in black and white on the page, rather than merely referred to by phrases such as 'it can be shown that.' You have been invited on a shared adventure of the mind—an exploration of spacetime in which careful reasoning is guided by graphical constructions and algebraic calculations. The most remarkable fact is that this adventure of the mind—of our powers of reasoning—corresponds to what is found 'out there' in the real physical spacetime that we inhabit. When we accelerate particles in linear colliders, the velocity *does* increase with time, just as was shown in the diagram on p.18. When hydrogen atoms fuse, energy *is* released in proportion to the change in rest mass. This set of connections—the fact that the basic physical workings of the universe make sense—is something that science has to take for granted. We have become accustomed to it, but it was by no

[1] Calaprice, A. (ed.), *The new quotable Einstein* (Princeton, NJ: Princeton University Press, 2005).

means obvious to early researchers. After all, plenty of human experience does not, on the face of it, make any sense. Early scientists made, by a step of faith, the conjecture that the world will make sense; and we make the same step, because when our studies of physics appear to be contradictory, we do not abandon hope but persevere until we can obtain an insight that makes the contradiction go away. The main postulates of Special Relativity are self-contradictory if one assumes that simultaneity is an absolute concept. Once we let go of that assumption, the main postulates make perfect sense. What convinces us that we are on the right track is not a proof that this way of thinking is best, but the sense of discovery, and the joy of discovery, when one crucial idea makes many others fall into place.

In Carl Sagan's science fiction story *Contact*, he introduces the idea that if a higher intelligence were at work in the universe, creating it or influencing it in some profound way, then perhaps there would be evidence in the form of a pattern where we would not expect one. He gave the example of a perfect circle appearing amongst the decimal digits of the number π, when they are written out in a rectangular array. This example is, unfortunately, rather a weak one. However, at least it makes an effort to suggest what sort of evidence an atheist such as Sagan might have found suggestive.

The remarkable intelligibility of the universe is something that we should not take for granted, but its origin lies outside anything we can discover by purely scientific methods. Perhaps it has no explanation that we can discover or understand. Or perhaps it has something in common with our experience of aesthetic beauty and moral pressure—it is a sign of a further reality that we can only engage with in ways that require active participation, not merely passive enquiry. In the huge and wonderful story of science we make the alarming discovery of a perfect circle in the fabric of the universe.

Index